旅
JOURNEY

思联设计的第一个20年之酒店及服务业

CL3's first 20 years' Journey on Hospitality design.

这里挑选的作品不是一个20年的回顾，而是会启发我们未来设计路程的动力。
Instead of a retrospective, these chosen projects represent inspirations for our future design directions.

目录 Content

思 THINK
计 PLAN
展 EXTEND
联 UNITE

序 — FOREWARD — 6
设计，此时此地 — DESIGN-THE HERE AND NOW — 10

思 — THINK — 18
东隅酒店 — EAST HOTEL — 24

计 — PLAN — 64
新加坡滨海湾金沙酒店 — MARINA BAY SANDS — 72

展 — EXTEND — 104
皇家太平洋酒店 — ROYAL PACIFIC HOTEL — 110
Mira Spa — MIRA SPA — 120
香格里拉酒店 — SHANGRI-LA HOTEL — 130

联 — UNITE — 142
唯港荟酒店 — HOTEL ICON — 148

活 LIVE

192 活
198 platform (1x2)
204 住宅项目

- 192 LIVE
- 198 PLATFPORM (1x2)
- 204 RESIDENCES

建 CONSTRUCT

218 建
224 时代地产商业综合会所
238 绿地北京大兴售楼处
244 仁恒置地广场
252 泰国度假小屋
260 泰国度假别墅

- 218 CONSTRUCT
- 224 TIMES CLUBHOUSE & COMMERCIAL COMPLEX
- 238 GREENLAND SALES OFFICE
- 244 YANLORD LANDMARK
- 252 TWIN VILLAS
- 260 BAN SURIYA HOUSE

东 EAST

268 东
274 上海依云水疗
280 西村日本料理
286 帘影楼
292 桂花楼
296 万科金域西岭
302 白沙源茶馆
308 华润无锡会所

- 268 EAST
- 274 EVIAN SPA
- 280 NISHIMURA
- 286 YING
- 292 GUI HUA LOU
- 296 VANKE SALES PAVILION
- 302 BAISHAYUAN TEAHOUSE
- 308 CR LAND CLUBHOUSE

感 APPRECIATE

314 感
322 北京香港马会会所
328 滩万日本料理
334 奥罗拉
340 FoFo
344 竞骏会
350 Adrenaline
354 香乐园

- 314 APPRECIATE
- 322 BEIJING HONG KONG JOCKEY CLUB CLUBHOUSE
- 328 NADAMAN
- 334 AURORA
- 340 FOFO
- 344 THE RACING CLUB
- 350 ADRENALINE
- 354 SHANG GARDEN

362 设计奖项
- 362 DESIGN AWARDS

378 团队
- 378 TEAM

序 刘小康
香港设计中心副主席

20世纪70年代,香港的现代设计崭露光芒。40年间,本地作品在亚洲、以至世界各地屡获好评,确立了其国际地位。香港设计师的创意不仅在世界舞台上发热发亮,更带动了正在不断改革开放的中国大陆市场,成就了各行各业的发展。

香港除了平面设计出色外,本地的室内设计也是另一个亮点。国家经济起飞,市场上涌现出大量房地产项目,从极尽奢华的豪宅,到以设计为先的商业大楼,各种各样,规模各异。客户有不同的期望和需要,室内设计业亦因此获得无限良机,庞大的祖国市场成为了香港室内设计师的试炼场。

本地室内设计能在内地独当一面原因有三:首先,香港设计师专业可靠,在注意每个细节之余,亦会考虑整体设计,更会管理项目,确保每一个项目都妥善地完成。再者,香港设计师对质量要求甚高,有助于提升内地项目的水平,尤其在酒店、商业大厦、住宅等项目方面,香港设计的质量不但能完全满足国际大型机构或企业的需要,更有助于吸引各地专才和游客到中国发展、旅游、投资、甚至居住。最后,香港室内设计,就如林伟而先生追求的东方精神那样,以现代的观点演绎东方美学的神髓。这正是香港设计鹤立鸡群的原因。

认识林伟而先生十多年,他的作品虽然低调,但处处显现着极高的质量。他在接受西方的教育的同时,亦能摄取中国文化元素,将其神韵呈现在作品之中,使他的创作既能让市场接受,又具国际视野,为年轻一辈树立典范。所谓的国际视野,是一种态度,一种尝试把中西文化融会贯通,既放眼国际亦不失自我,并且在宏观世界亦拥有自己独特的文化理念的专业态度。

刘小康

刘小康，香港著名设计师及艺术家，靳与刘设计有限公司创办人，从事设计、公共艺术创作及雕塑创作等，作品获多个博物馆珍藏；同时参与艺术教育和推广，担任多处非赢利设计机构的领导职位，包括香港设计中心董事局副主席及北京歌华创意中心总监等。2006年，刘小康获香港特别行政区政府颁授铜紫荆星章，以肯定其在国际舞台上为提升香港设计形象所付出的努力。

提及林伟而先生的作品，就不能不提他的"彩灯大观园"。作品举世瞩目，于2003年的非典型肺炎(SARS)过后面世。这巨型的花灯结构在紧迫的时间、恶劣的天气、复杂的技术问题之中顺利地完成，正展示出香港人的团结精神和坚毅气魄！林伟而先生在作品中巧妙地揉合了现代建筑与传统竹结构工艺，打造出东方的美态。这彩灯包含了许多中国文化艺术元素和传统景致，而且功能上既是一个展览馆，亦是一项大型公共艺术作品，正体现了香港在中国现代设计舞台所提出的方向和标杆。

唯港荟酒店是林伟而先生为香港打造的另一新地标。这个项目有完整而清晰的设计方向，在建筑、室内设计和艺术等范畴都发挥得淋漓尽致。今天唯港荟吸引着世界各地的旅客，其受欢迎程度可以媲美五六星级酒店，是炙手可热的时尚选择。尽管没有天文数字的财政支持，林伟而先生凭借其创意，设计出如此成功的一座艺术品，不止是本地同业的典范，内地的不同行业也纷纷研究此项目的成功之道。唯港荟体现了香港设计师对整体设计和艺术价值的热切追求，我相信我们一直沿着这个方向前进，香港设计必定能再创佳绩，成就更多更富创意、质量更高的作品。

林伟而先生品味高雅，而他把东方文化融入作品的风格也正是其魅力所在，加上团队创新专业的态度，成就了思联光辉的二十载！

恭喜思联走过这二十个寒暑，祝你们鹏程万里，向更多个更成功的二十年进发！

Foreword by Freeman Lau
Vice Chairman, Hong Kong Design Centre

Hong Kong's design industry started in the early 1970's. Within a period of 40 years, the works of Hong Kong designers have gained a respected status in Asia and all over the world, receiving important international recognition. Not only is the creativity of Hong Kong designers known on the world stage, it is also affecting the developing market in Mainland China and has contributed to the growth of various design industries.

In addition to graphic design, interior design is one of Hong Kong's great exports. With the burgeoning Mainland Chinese economy, there is great demand from real estate developers for design for everything from high-end residences and well-appointed commercial buildings to all sorts of projects of different scale and usage. Every client has different expectations and requirements, providing interior designers unlimited opportunities to exercise their talents. The China market has also become a major testing ground for Hong Kong interior designers.

There are three main reasons why Hong Kong interior designers are so well-regarded in China. Firstly, their reliability and professionalism, in paying attention to every detail, as well as taking into account the commercial viability of the designs. In addition, Hong Kong interior designers are conscientious in the running of projects and strive to ensure satisfactory hand-over to the client. Thirdly, Hong Kong designers have high expectations in terms of quality control, which had helped raise the general standards of interior design in China. In the fields of hotel, commercial and residential design, the quality of Hong Kong interior design not only satisfies the needs of multi-national corporations and enterprises, it also helps attract professionals and visitors to China for work, travel, investment or even immigration. Lastly, Hong Kong designers such as William are familiar with the spirit of Eastern design, and are good at giving Eastern aesthetics a modern interpretation, which makes them stand out from the crowd.

I have known William for more than 10 years. Even though his works are often low-key, they always represent high standards. Although educated in the West, William has an excellent grasp of Chinese culture, and incorporates its essence into his work, and in the process has received wide acceptance among his peers. William's designs have true international vision, and are inspirational to the younger generation of designers. This international vision is an attitude that bridges the East and the West. Keen to understand the world around us while maintaining a sense of self, he maintains a professional attitude that looks to the world while staying true to one's core cultural heritage.

Freeman Lau Siu Hong

Being a renowned Hong Kong designer, Freeman Lau is Chairman in Kan & Lau Design Consultants. Apart from designing, Freeman works on public art and sculptures. His artworks and sculptures are collected by museums all over the world. Currently, Freeman devotes himself to art education and promotion. He takes leading roles in many non-profit making organizations, including vice-chairman of the Board of Directors of Hong Kong Design Centre and director of Beijing Creative Centre. In 2006, Freeman was awarded a Bronze Bauhinia Star and the effort he paid for enhancing the image of Hong Kong design was appreciated.

In discussing William's works, one thinks of Lantern Wonderland, which went on CNN when it appeared in 2003, right after Hong Kong's SARS epidemic. This enormous lantern-shaped structure was built on a tight schedule in inclement weather, and the construction team had to overcome various technical difficulties to complete the project on time, which showed the resilience of the people of Hong Kong and their ability to come together during hard times. In this work, William skillfully combined contemporary architectural technology with traditional bamboo construction to create a structure that is a perfect expression of Eastern beauty. The Lantern Wonderland incorporated elements of Chinese art and culture as well as traditional visual motifs; this piece epitomizes the direction and standards that Hong Kong can play in the field of design in China today.

Hotel ICON is literally by now a Hong Kong icon. Designed by William, this project has a clear and coherent design vision, which found perfect expression in architecture, interior design and art. Today, Hotel ICON has achieved success by attracting guests from all over the world, and its popularity is on par with the top hotels in Hong Kong. With a relatively modest budget, William was able to use his ingenuity to deliver good design that is also commercially successful. Hotel ICON's success has become a case study for designers and hotel professionals locally and in greater China as well. It embodies Hong Kong interior designers' commitment to good design and artistic value. I am hopeful with projects like this, Hong Kong design will keep on reaching new heights, and there will be more creative and high quality works produced in the future.

William Lim has masterfully incorporated Eastern culture into his design which has shaped CL3's distinct aesthetics. With the professionalism and the innovation of his team members it is no surprise that their projects have been receiving awards year after year for the past 20 years.

Congratulations to CL3 on its 20th anniversary. I wish them all the best and hope there will be even greater success for CL3 in the next 20 years!

设计，此时此地

琚: 琚宾　　**林**: 林伟而　　**何**: 何宗宪

　　本次颇有趣味的自由对话发生在林伟而先生的办公室中。与见诸传统媒体中的访谈不同的是，这里没有"记者"这个角色的存在，林伟而、琚宾和何宗宪借对《旅》一书的探讨为契机，以艺术与设计、设计流程、设计与教育以及设计生涯这些零散的关注点为主题进行交流，并未试图得出某些确定的结论，而是以设计师对专业的敏感，对设计中面临的问题加以反思，期待以此让设计的使命更加清晰。

琚：艺术与建筑具备很强的精神性，但因为室内设计具有辐射到生活的末端的属性，艺术性相对弱一些。应该如何在跨界的设计中实现身份的转换，对艺术形成一个理解的方法？在设计中，除了豪华这种常见的手法外，设计师还能够提供什么？

林：我是50年代的人，读书是在70年代。那个时期，早期现代主义风格的设计比较多，当时完全脱离了奢华的语境，奢华只是一个符号。我在设计中更追求空间的感觉而不是奢华的装饰。例如密斯凡德罗的巴塞罗那德国馆，使用了很多奢华的材料，如大理石和红丝绒，但是空间的感觉还是强于材料的奢华感。所以我认为做设计空间的感觉更重要，还是要追求内容和气氛，可以采用装饰达到这个气氛，奢华的背后要有内容。

琚：例如刘易斯·巴拉甘，表达了一种宁静和内向化的精神。

林：巴拉甘使用了最普通的材料来表现空间的精神，但是他的建筑中对水的控制显然不是一般的家庭可以承受的，我想这可能就表达了建筑、艺术与奢华之间的关系。

琚：在我看来在艺术中悲情的部分更容易感动人，而艺术是隐形地表达当下社会问题的，您在香港也做了很多的艺术和装置作品，那么您是如何看待您的艺术作品的？

林：在我越来越接近艺术的时候，我发现艺术家的思维和建筑师的思维是有很大不同的。艺术家认为建筑师了不起的地方是，建筑师能够实现很多大型项目，对环境、艺术和空间的想法能够有成品表现出来。而我认为艺术家做设计通常会追求一种美感。设计从来不会追求"悲情"。我做装置是希望有所突破和挑战，希望能够把人的其他感觉融入其中。例如丹尼尔·李博斯金德的柏林犹太人纪念馆，完全打破了传统的美感，融入了很多沉重的精神感觉，跨越了纯粹的建筑与室内的设计，融入了人的情感。

何：我认为应当让生活的艺术接近设计师的创作，不需要特意将激烈的感情投入其中，而是让人们参与、欣赏和品鉴。

林：安尼诗·卡普尔近日在芝加哥完成了一个不锈钢的艺术装置，我看过之后非常感动。这个装置与不同年龄的人之间产生了不同的反应，并且大家都很热烈很直接地参与到这个装置中去，成为这个作品的一部分。这个作品就拥有着很强的艺术性和社会性，给人们带来了很多欢乐。艺术家的作品能够从不同角度使人们参与和欣赏，如果设计能够做到这点就是非常好的了。

琚：例如您的"鱼"这个装置作品，最让我感动的是，您在拍老百姓看鱼时候的状态，而不是装置作品本身。这其中体现的艺术与设计、与社会搭接的关系，让人产生思考。那么，香港的社会是一种什么样的文化生态，能够允许您愿意进行这样的实践？

林：其实我最早做装置是在2003年，也就是灯笼的装置，当时香港旅游发展局希望带动旅游项目。装置完成后，反应热烈，来参观的人非常多，很多都是老百姓，对他们来说这是一件非常新鲜的事情。对我来说，做设计给更多人看非常有趣，只要有机会我就会去做。香港地方比较小，有兴趣做的人也少，因此相对来说机会就比较多。每两年会有威尼斯双年展和香港深圳建筑艺术双年展，以及Detour，因此我们经常有这样的机会争取做一些有趣的项目。其实这些项目构思并不难，并且最终的成果是由很多我们的同侪来实现的。

琚：我在做设计的时候，因为多种因素的综合博弈，会有很多的无奈和妥协，可否给我们这些年轻人一些建议，在设计的思考和策略方面？

何：我一定要令自己快乐才能做设计。把让我痛苦的部分全部删掉，我就会比较快乐。

林：每一个设计都有开心和痛苦的时候，就像人生一样。我们即使很小心地选择项目，还是会有令人痛苦的东西出现，完全不能避免，关键是出现问题的时候如何解决。设计做得出色，大家都会很开心。例如东隅酒店，甲方对我们很支持，我们会争取做好每一件事情，把甲方的想法体现在设计里。对于甲方来说，他们其实希望设计师发挥他们最好的潜力来做事情。我们工作很多年了，不会把公司的问题延伸到私人生活里。我会和甲方沟通，一起努力解决事情，但不是一定要在今晚熬夜解决它。我们的合约里有约定，如果甲方对我们的要求很不合理，我们会终止合约。

琚：在出现问题的时候，如何来解决呢？

林：最好的做法是和甲方一起想出解决的方法，让他知道你在用心解决问题，只是需要一点时间。我会倾向于用运动解压。我们也经历过很多要求不合理的甲方。经过了二十年，我们也在慢慢学习如何选择甲方。

琚：您和公司员工之间，除了项目本身的设计策略和设计方法外，还会单独讨论设计思想么？

林：很多老员工，不需要讲也会了解，我们会一起研究，做出一个新的概念。

何：您后期的作品开始趋向于一种独特的气质，出现一些很难解读的感觉。

琚：我认为林伟而先生抓住了他的设计作品中内在的灵魂。

林：每个作品都有令人遗憾的地方，我们希望未来的项目能够让我们更满意。我在实践中得到的经验让我能够解决很多问题，但是创意是不应该凭借以往经验的，应该是往前推进的。

琚：创意是一把双刃剑。

林：甲方请设计师做新项目，设计师是否会重复自己的创意？这确实是一个问题。

琚：您是否有一个方法，能够让自己的设计一直保持新鲜和旺盛？我很担心自己对设计会慢慢失去热情，我希望自己越来越热爱设计，设计需要呵护，需要屏蔽掉很多东西，独立于社会。

林：就像一个蜡烛，在合适的环境中会燃烧到底，若在风大的地方，很快就熄灭了。人对设计的热情，也需要外界的支持和认可。甲方和设计师一起成长，如同一个有着良好环境的蜡烛；但是如果他不断打击你的设计欲望，设计师就很难保持热情的心态。很多优秀的设计师因为一个冲击就放弃了，非常可惜。我认为一是要对这个行业有热爱，二是要小心呵护自己的创作热情。

何：我非常羡慕林伟而通过这么多年的坚持达到了很自然的设计状态。保持了设计的感觉，没有被艺术吞没或者掩盖。

琚：并且他的设计有了突破性的形态关系，对历史与设计的关系控制得很巧妙，在气质上保持了一致。这不是一朝一夕能够达到的。很多年轻的设计师过于追求设计的震撼效果，但是经历和思考却不够，使得他的设计损失了很多细节。所以我认为应当告诉年轻的设计师们，不要在尚未掌握的时候一味地模仿形式。

林：我觉得设计师应当慢慢地沉淀下来。起步的十年还是非常艰难的，但是幸运的是很多客户给了我们机会，例如Nike找到我们的时候，我们的公司只有两三年的历史。设计不能操之过急，而是要慢慢研究慢慢吸收。设计的精神是空间，人在空间里的感受。很多年轻设计师会用很好的照片参加各种竞赛，但是二维的照片只有一个角度，而空间应该从立体的角度观察，如果这样还是能够感到震撼，那么这就是一个好设计。

琚：在教育方面，您有什么看法呢？

林：在这方面我不是特别了解，但是我觉得即使是在香港，和国外也有一点不同：在上课的时候，大部分学生都比较被动，不够活跃。建筑设计很大一部分工作需要沟通，我们的沟通方式里，除了图像，语言也是很重要的一部分，说出来并让别人接受我们的想法。还有一个问题，国内外可能都很普遍，电脑的普及使大家太容易做出一个震撼的效果，那么老师就要很清楚，在软件做出的效果里看出空间的精神，教育学生表达空间而不是做一个漂亮的效果图。

琚：我非常有同感，在教育院校带学生做毕业设计的时候，很多学生都沉溺在软件带来的形式中。但是技能是不需要教的，要给学生思想。一线设计师到大学里代课也是一把双刃剑，若设计师自己还没有想明白，就不应该去教学生。这应当是个非常慎重的事情。您教设计方法多一些，还是思想交流多一些？

林：我是一定要教设计方法的。不会因为他是研究生我就假定他有经验，我会从一个基础上开始，很快地发现他们的基础在哪里，之后因材施教。我希望能够引导学生在他们的设计中找到一个灵魂，之后再延续成为一个设计。如果设计没有灵魂，即使形式很震撼，也是一个空洞的设计。

何：有位作家曾说过，人生如茶。我觉得您的设计，第一道茶很浓，因为您在设计中融入了很多思考；第二道是香，作品很贴近生活，给人一种宁静的体验，享受生活；第三道则是淡，是甘，是一种东方的特质。

琚：我觉得从人的角度看，您先是技高人智，之后品高人贤，现在则是德高人善了。您对设计的感悟让我很受益。我在读过《藏物论》、《闲情偶寄》、《园冶》等中国传统书籍后，开始更深地理解中国人的生活方式。不是旧式那种标榜的浮华美，而是内在的文化气质，其更能代表中国的特质，更符合东方人的思考。

何：在建筑师中，我受库哈斯的影响是非常大的。

林：我受库哈斯的影响也是非常大的，他给了我很多启发，他的建筑要走进去才能够体验和感受空间。还有密斯安德罗的设计，让我对空间的观感有了很多理解。卡罗斯卡帕也是我非常欣赏的建筑大师。

琚：我在看过很多现代建筑大师的作品之后，最后看到柯布西耶，才发现，很多建筑师都是跟随柯布西耶的体系的。柯布西耶几乎是一个为建筑而来的人，很难超越。没有材料的变化很难诞生更有影响力的大师。

林：一个很有趣的地方是，柯布西耶把结构和楼板、外墙分开，这种概念其实是中国很早以前就有的，这样功能不需要受到结构的限制。

琚：让我很感动的一个建筑是柯布西耶给母亲做的家，设计了很多有趣的细节，例如给猫和狗的阳台，让我很受触动。国内的建筑师很多时候并没有那么注重生活细节。

何：路易斯康也是一个非常有代表性的人物。

林：他的设计非常有东方的特质。

琚：还有一个很有趣的细节，康在完成建筑后，请巴拉甘来一起进行讨论。而园林最终的形成，得益于巴拉甘的建议。

林：OMA日前刚完成了康乃尔大学建筑系馆的设计，概念是非常领先的。

琚：柯布西耶说，设计应是一种信仰。我们要在社会中保持自我。当我走进大师的作品中的时候，就会有一种愉悦感，就会告诉我自己，我一定要自己完成设计，连桌子都不能交给别人做！这种感觉就像打了兴奋剂一样。

何：我同意你的说法，设计有时候如同一种宗教。

Design-The here and now

JB: JuBin   WL: William Lim   JH: Joey Ho

This interesting and free ranging conversation took place in WL's office. Unlike traditional media interviews, there were no "journalists" present. Rather, WL, JB and JH used the book *Journey* as a starting point to explore such diverse topics as art and design, the design process, design and education as well as design and life. The point of the discussion is not so much to draw definite conclusions about these topics, but rather employ the sensitivities of professional designers to reflect upon these issues as they encounter them in their work, and in the process hopefully clarify their design missions.

JB: Architecture and art seem to have a strong self ideality, which is not easily present in interior design. How can we obtain a change of identity from across disciplines so as to achieve a dialogue with art? Aside from sumptuous spaces, what else can interior designers provide?

WL: I'm from the 1950s, and went to school in the 1970s. At that time, Modernism was architecture's driving force, and luxury, which was just a symbol, was not part of the design vocabulary at all. In my designs, I aim towards conjuring feelings with spaces rather than pursuing elaborate decorations. For example, Mies van der Rohe's Barcelona Pavilion incorporated lavish materials such as marbles and velvet, but the materiality is only a reinforcement and refinement of his sense of space. I think in interior design, spatial atmosphere is most important, and one has to look for a strong concept. You can use materiality to enhance the atmosphere, but behind all this, there should be a clean concept.

JB: Architects like Luis Barragán expressed a tranquil and internalized spirit in his works.

WL: Barragán used the most common materials to portray the spirit of spaces, but his control of water in his architecture is not something that can be achieved in ordinary homes. I guess this expresses the relationship between architecture, art and luxury.

JB: I think the power art gives to people is often ultimate and tragic. You have done a lot of art installations in Hong Kong. How do you view your own artistic work?

WL: As I get deeper into art, I find that the way artists think and the way architects think are vastly different. Artists think that the marvelous thing about architects is that they can realize many large projects, and express their thoughts about the environment, art and space in their works. I think that most interior design seek a kind of spatial "beauty". It is never about other aspects of emotions like sadness or unease. I work with installations because I am looking for some sort of breakthrough and challenge, and try to incorporate other kinds of human emotions into my art. For example, Daniel Libeskind's Jewish Museum in Berlin went against traditional ideas of beauty and incorporated some somber and suppressed spaces, which went beyond mere architecture and interior design by letting the viewer experience other human emotions.

JH: I think we should allow the art of everyday living to become intertwined with designers' works, and do not need to infuse our works with intense emotions. Rather, we ought to allow people to participate, enjoy and critique our works.

WL: I recently saw the "bean", a public art by Arish Kapoor in Chicago's Millennium Park, which greatly moved me. The work allowed people of different ages to interact and have different reactions, and people participated in the artwork enthusiastically. This piece of art has a strong artistic and social component, and brings people a great deal of pleasure. Works of art can be appreciated from different angles, and it would be good if interior design is capable of this.

JB: What touched me most about your installation work "Lantern Wonderland 2011" is that you record the reactions of people as they look at your work, rather than the installation itself. This thought-provoking work combines both art and architecture, and their interconnection with society. What kind of cultural scene does Hong Kong possess that allows you to do this?

WL: I began doing installations in 2003, and started with the lantern installation. It was after SARS, and the Hong Kong Tourism Board wanted to bring back tourism to Hong Kong. When the installation was completed, it attracted 150,000 visitors, and the response was very positive. Most of the visitors were grassroots people, and to them the installation is easy to understand yet magical. To me, exposing my designs to a wider audience is an interesting experience, and I would like to do more if I get the chance. Hong Kong is a small place, and there are few people interested in public art projects, so for me there are more opportunities. The Venice Architectural Biennale, the Shenzhen-Hong Kong Bi-City Biennale of Urbanism and Architecture, as well as Detour, are some of the opportunities to do installation projects. Coming up with the concepts for these projects is not difficult, and I work with skilled people in various trades to have them realized.

JB: Nowadays, I run into a lot of difficulties with designing. Can you give some advice to young people like me on ways of thinking and strategizing?

JH: I can only do design if I'm happy. If you take away all the elements in the design process that are aggravating, then I'd be happier.

WL: In every project there are things that are painful and things that make me happy. It's just like life itself. Even if we were careful in picking our projects, difficulties are bound to crop up-they are unavoidable. The important thing is how to deal with problems that appear. In fact, everyone wants design to be done well, and the client really hopes that designers will realize their true potential. For example, with East Hotel, the client was very supportive of us, and we would fight for the best results while incorporating the client's ideas. I have been in the business for long enough to know not to bring worries home. I would communicate with clients and try to solve any problems that come up, but I realize that not all problems can be solved right away. We have a clause in our contract that if the client's requests are too unreasonable, we have the right to terminate the agreement.

JB: What do you do when problems appear?

WL: The best way is to work with the client to come up with a solution. You have to let the client know that you are doing your best to solve the problem, but you just need some time. I have also encountered many unreasonable clients. Our company has been around for 20 years, and we are slowly learning how to pick our clients.

JB: Do you discuss design philosophy with your employees?

WL: I have been with the same team for many years, and they understand my thinking and what CL3 stands for. For every new project, we would brainstorm together to come up with new ideas.

JH: Your recent works have tended towards a unique style, and conjures up an indefinable feeling.

JB: I think WL has grasped the soul of his work.

WL: In every project, there are things that I regret. I hope in the future, there will be projects we'd be happy with. With practice and experience I can come up with many ways to solve problems. Yet creativity should not depend on past experiences, but ought to keep moving forward.

JB: Creativity is a double edged sword.

WL: The client hires a designer to do a new project. Does the designer repeat his or her creativity? That is a big question.

JB: Do you have ways to keep your creativity fresh and strong? I am worried that I will gradually lose my passion for design. I hope I will love design more and more. Design needs to be nurtured, it should transcend our society.

WL: Creativity is like a candle. In the right environment it will keep burning to the end, but in a windy place, it will soon get blown out. A designer's enthusiasm towards design needs support and reinforcement. If the client support the designer, that is like the candle being allowed to burn in a good environment. However, if the client constantly attacks your desire to create, it is hard for designers to maintain their passion. Many good designers give up because they cannot withstand such setbacks, which is a great pity. I think first of all you have to be passionate about this industry, and secondly carefully nurture your creative passion.

JH: I admire William for maintaining a design style all these years, that seems to come so naturally, as if it's a part of himself, and which did not get swallowed up or covered up by his artistic side.

JB: His design seems to have achieved a breakthrough. He has mastered control of the relationship between tradition and innovation, and has sustained a consistent quality. This is not something that can be achieved in a short period of time. Many young designers are fond of trying to create awesome effects, but they don't have sufficient experience nor do they apply adequate thought into their designs, so their designs lack details. I think one should tell young designers that they should have proper grasp of design instead of trying to chase trends.

WL: I think designers need to slowly develop; the first 10 years was very tough for CL3, but fortunately many clients offered us opportunities to learn. For example, when Nike hired us, our company had only been established for one to two years. In design you cannot rush things, but have to slowly learn and absorb. The essence of interior design is space and the feelings it inspires in people. Many young designers would take pretty pictures of their projects and enter them into competitions, but two dimensional photographs only sees things from one angle. Space should be experienced in three-dimensions. If a design is able to make an impression within three-dimensional space, then it's a good design.

JB: What do you think about interior design education?

WL: I am not too familiar with that, but I think interior design education in Hong Kong is somewhat different than that in other countries. Most Hong Kong students are too passive and do not participate enough. Architecture and design is very much about communication. In addition to drawings, it is important that we communicate with language. Another problem is probably common in both Hong Kong and the rest of the world. The wide-spread use of computer means that it is relatively easy to create a striking visual effect. Teachers must make it clear to their students that beyond the effects, one should be able to observe the concept behind the space. They should teach student that spatial design is not just about creating impressive renderings.

JB: I very much agree with you. In some education academies, graduating students are overly obsessed with the software rendering. However, technique does not need to be taught. Rather, students should be given room to think. Guest lectures by first-rate designers can be good and bad thing. If the designer has not thought through certain issues, then he or she should not be teaching students, for education is a very serious matter. When you teach, do you emphasize design methods or exchange of ideas?

WL: I always teach design methods. I would start from the basics. I don't assume students already have experience, even if they are Master's students. After finding out the students' capability, I would teach according to their abilities. I hope I can help students find the soul in their design, and from that extend into their work. If a design has no soul, it is empty, no matter how impressive it looks.

JH: A writer once said that life is like tasting tea. I think it can be applied to William's design. The first cup is strong, because you incorporated a great deal of thinking into it. The second cup is fragrant, because your work is closely related to daily living, and gives people the chance for quiet contemplation. The third cup is mild with a lasting taste, with a distinctly Asian, humble quality.

JB: From a human perspective, it can be said that you start off having wisdom, then character, and finally virtue. Your understanding of design has really inspired me. The designer Ye Fang recently recommended some books to me, such as *On Creation and Leisurely Life*, which made me understand the Chinese way of life, the essence of which is not the sort of extravagance popularized in the Qing dynasty, but a cultural temperament upheld during the Ming dynasty. The latter is more representative of the Chinese character and closer to Chinese thinking.

JH: Among architects, I was most influenced by Rem Koolhaas.

WL: So was I. He inspired me a great deal. With his work, you have to step into the space to be able to experience its impact. Also Mies van der Rohe's designs gave me a new understanding about space. I also admire Carlo Scarpa a great deal.

JB: I studied a lot of works by modern architects, and when at last I came to Le Corbusier, I discovered that many architects follow him. Le Corbusier lived almost entirely for architecture, and it is hard to surpass him. Unless we come up with new building materials, it would be difficult for the world to produce a more influential master.

WL: An interesting point is that Le Corbusier made a separation between structure and enclosures, but this concept has long existed in China. This allowed function to be unhampered by structure.

JB: I was really touched by the house that Le Corbusier designed for his mother, in which he incorporated many interesting details, like a balcony for cats and dogs. In China, not many designers understand the art of living.

JH: Louis Khan is also a very representative figure.

WL: His designs have a distinctly Asian flavor.

JB: There is a very interesting anecdote. After Khan finished a building, he invited Luis Barragán to look at the landscaping. It was only when Barragán approved that Khan became satisfied with his own work.

JB: Le Corbusier said that design should be a religion. We have to stay true to ourselves in society. When I step into a building designed by a master, I feel a sense of joy. I tell myself that I have to remain in design. I can't let other people design for me--I have to design even my own desk. This feeling is exhilarating.

JH: I agree with you. Sometimes design can be like a religion.

何宗宪

生于台湾、在新加坡长大的香港建筑及室内设计师何宗宪于香港大学修毕建筑硕士，随后于2002年凭借其用之不竭的活力及热情自行创立了 **Joey Ho Design Limited** 设计公司。设计项目类型涵盖广泛，屡获殊荣，在海外及本地的设计项目中，众多知名作品至今已获逾90个奖项。

何宗宪与设计团队不断探索环境设计的新理念。他们不仅探索建筑如何与现代生活紧扣在一起，而且探讨建筑如何与各种系统相连结。在室内的设计风格上是以他们的触觉，追求朴实、简洁的空间，在实用空间及概念之间取其平衡点，营造和谐的新时尚生活环境。

他们的风格以当代精神见称，现代创新与优雅实用并存。作品类型及规模广泛，从展览装置到精致的室内布局等，均秉持设计映现一种透过日常境况去塑造生活的精神。

琚宾

HSD水平线空间设计|北京|深圳|首席创意总监，就读于中央美术学院，高级建筑室内设计师，中央美术学院建筑学院、清华大学美术学院设计实践导师。注重设计文化的多元与共生，善于运用传统文化的精神，设计手法当代却不失传统。他的设计作品，将"当代性"、"文化性"、"艺术性"共溶、共生，以此作为设计语言用于空间表达。

从传统与当下的共通、碰撞处，找寻设计的灵感；在艺术与生活的交错、和谐处，追求设计的本质。

他喜欢在历史的记忆碎片与当下思想的结合中，寻找设计文化的精神诉求。在他的设计中，不断追求的创新精神，也让他在社会上和行业内获得诸多好评。一些与他一样具有设计梦想的年轻人加入到了由他带领的HSD水平线设计团队。HSD团队主张：无创新，不设计。

Joey Ho

Joey Ho draws his creative inspiration from the far-reaching corners of Asia. Born in Taiwan, raised in Singapore and gaining his Masters in Architecture from the University of Hong Kong, each of these culturally diverse yet artistically vibrant qualities have played their part in fashioning Joey's unique and avant-garde perspective of the world.

Planting his roots in Hong Kong, Joey set up Joey Ho Design Limited in 2002, attracting a young and energetic team of hugely talented individuals. Their diverse skills, styles and disciplines come together to create dynamic and engaging spaces where people can thrive and blossom. We believe everyone is influenced and inspired by their architectural surroundings and with each new project comes the opportunity to help improve their environment and standard of living. To-date, Joey's designs have garnered more than 90 internationally recognized awards.

JuBin

Chief Creative Director in Horizontal Space Design Ltd., Corp. of Shenzhen and Beijing, Graduated from China Central Academy of Fine Arts, Senior Construction Interior Designer. Mentoring in Design Practices to Architecture school of China Central Academy of Fine Arts, and Academy of Fine Arts of Tsinghua University. Focusing on the pluralism and symbiosis of the Design Culture, he is dexterous in embodying the essence of traditional culture into his designs, making them modern but also with the quintessence of tradition.

In his design works, Contemporaneity, culture and artistry were emerged and made intergrowth in his works, which was employed as the design language in his spatial expression. He seeks the design aspiration through the similarities and collisions under circumstances of the tradition and contemporary, chasing for the essence of design among the intersection and harmony of life and art.

He enjoys looking for the spiritual demand of designing culture through the combination of historical and contemporary thoughts. The innovative spirit in his design wins him a good reputation in both the society and the field. Young men with the same designing dreams as him joined his team - HSD. HSD Claims: No design can be achieved without innovation.

思
THINK

思
四年半的思路

思，是做每件事的开始。早上醒来，是否起床，已经是第一所思，所以每天我们都做着无数的思考。胡思乱想，可能是创作最好的开始。

每个项目，犹如一个人、一件物品，都需要有它的个性，个性强的就可以像广东话的"出位"，而这个性与风格无关。任何一种风格只要做得好，都可以是好的设计，可以表达一种很强的个性，而某些设计师更擅长某种风格。以电影导演为例：一个好的导演，他用古装、时装都可以拍出具有自己独特个性的电影，为什么设计师不可以？

在开始一个设计的时候，CL3不会预先给它定位，让思路广阔，多去幻想，从而探索一些新的设计思路。我们会把这些想法与客户沟通，看对方能否接受，再慢慢地由浅入深，由广到精，在项目中的每个细节展示设计概念。

东隅酒店的设计之初，建筑设计没有定案，客户明白一脉贯通可以使酒店有清晰明确的风格，并且用室内设计先行带动，可以先考虑客人的感官和感受，然后再反映在建筑上，所以委托CL3在酒店开业四年半前介入，负责酒店的全部室内设计。当时酒店也尚未命名，没有风格上的限制，只知道该酒店是一个位于香港传统的住宅区的商务酒店。于是我们决定先从客房设计入手，因为它重复性很大，而且直接影响结构、管井等。酒店若要有商业竞争力，房间就不能太大，客户已定了26平米。我们的切入点是考虑如何利用26平米，来满足一个商务旅客的各种需求。我们用自己商务出差的经历来推敲每个细节：出差时行李箱多大？是否会把衣物放进衣柜？是否用书桌？是否泡浴？所有这些经历都融入房间设计里。我们在淋浴时，最讨厌的是打开水龙头时冷水淋在身上，所以我们把花洒开关放在入口的地方，先开水再步入淋浴间。通过这样对每一个细节的推敲，最后用了约两年的时间，把房间布局调整成现在的样子。我们希望客人一早醒来已知道身在何处，所以用了一些东方元素和一些香港的照片。就这样，风格也慢慢明确了下来。

一般酒店可能会用不同的设计师负责不同的区域：房间、公共部分或餐饮部分，CL3却有幸被委托设计整座酒店。在设计房间的同时，我们也在思考其它部分的设计，以客人感受为主的探索思路推敲每一个细节。酒店的公共部分是一些长而有细微弧度的建筑空间，我们尊重空间的特点及它与外面景观的关系，来设计所有空间的功能布局。我们很大胆地利用几种很简单朴素的材料，布置整个酒店空间，并大量使用浅色的工业木材，线条也极简约，形成一个统一纯净的空间，刚好与公关公司建议的中文名字"东隅"呼应。

每一个好的项目都应由客户作明确的定位后，委托设计师从中勾勒出一个理念，在设计中融汇贯通，从方案开始至施工权利配合，把理念渗透到每个细节。而在大型项目或包含多种功能的酒店项目中，这一点尤为重要。

设计把握理念，把握思路，其它所有的配套如建筑、结构、机电、灯光等都要跟随设计的方向。东隅酒店在大堂设计上坚持了这一点。我们希望大堂有两个主要元素，一是接待台，二是楼梯。这两种元素代表了香港的两个极端：阴阳的对比。在接待台的设计中，我们采用了一根16米长，从结构柱上支撑起来的悬挑的发光体，营造出了一种自然的东方美，给人宁静的沉淀，但同时又为人带来惊喜。同样地，楼梯的设计是要表达杂乱而富有动感的城市魅力，所以采用了不规则的钢结构及玻璃的梯级。幸得客户的支持，我们特意为此设计做了一个1比5的模型，又搭建了1比1的局部来研究所有的节点。

一个四年半的项目，要坚持一个清晰的思路，也要坚持克服一切工作上的困难。在最关键的时刻，CL3的项目经理离职，幸好有团队一致的坚持，项目终于成功地完成。东隅酒店是我们第一个全面设计的酒店项目，在此我们要感谢太古集团的支持及信任。

Think
A thought-provoking journey of four and a half years

All action starts with a conscious decision. You wake up in the morning, and your first thought is whether to get out of bed. Every day, countless thoughts pass through our heads. Letting your thoughts wander may be the best way to begin the creative process.

When starting a design project, William tries not to give it too much definition or emphasize a particular style. He thinks every project, like a person or an item, must have its own character. Those with strong characters are said to stand out, yet this character has nothing to do with style.

We believe that any style, if done well, can be good design and express strong character, and that some designers are better at certain styles. Take film directing as an example: a good film director is able to make films with his own characteristics no matter whether he is making an historical epic or a modern drama. So why shouldn't designers do the same?

At the start of a project CL3 would brainstorm and let our imaginations run free as we start to explore new design concepts. We would then present these new ideas to our clients and hope that they share our views. From the initial idea we would dig deeper and go from the general to the specific, applying the design concepts to every detail within the project.

When CL3 worked on the design of EAST, we started initially with the guest rooms. At that time, the architecture was still on the drawing board, and the client realized that if the project began with interior design, the design could take into account the internal functions and its relationship to view, etc, which will then be reflected in the architecture. That was why CL3 was taken on board four and a half years before the hotel's official opening. We began with the guest rooms because they are the most important part of a hotel, and also because they are repeated 300 times directly affecting the structure and the systems design. When we began the design, the hotel did not yet have a name, and we were not encumbered by any pre-conceived styles. We only knew that the hotel was to be a business hotel located in a traditionally residential Hong Kong neighborhood. In order for it to be commercially viable, the guest rooms could not be too large, with the client stipulating that they be around 26 square meters.

CL3's foremost concern was how to make the best use of the 26 square meters to satisfy the needs of business travelers. We applied our own experience traveling for business to determine every detail of the guest rooms. For example, how large are our suitcases? Do we unpack our luggage and store our clothes in the dresser? How often do we use the writing desk? Do we take a bath? All these experiences were incorporated into the design of each guest room. When taking a shower, it is not pleasant being splashed with cold water, so we put the water faucet close to the entrance of the shower stall so that guests can turn the water on before stepping in. Paying attention to every detail, we spent two years fine-tuning the rooms. We wanted the guests to know where they are upon waking up in the morning, so we included a photo mural of Hong Kong and, of course, the incredible view helps. In this way, the character of the hotel became more and more defined.

As we worked on the guest rooms, we also considered other parts of the hotel at the same time. Usually, hotels would employ different designers for different areas, like guest rooms, public areas and food and beverage. We were lucky in that the client realized having consistency allowed the hotel to have a strong and clear identity, and we were given the responsibility of designing the entire hotel. We extend the same thinking process from the rooms to the public areas. We respected the architecture and the spaces it enclosed, the views it framed, etc in determining the functional layout of the hotel.

We boldly utilized just a few simple materials that are applied in every part of the hotel so as to create a pure and unified whole. Our choice of materials was very understated, two shades of engineered veneer, limestone, glass and stainless steel. The reception desk is a 16 meter-long illuminated beam cantilevered from a single column, which exudes a natural Oriental beauty.

CL3 believes that every project should start with a clear vision, from which the designer should then come up with concepts or ideas, which would then be carried through the entire project like the project's DNA, the concept should permeate every detail. This is all the more important for large-scale projects, or multi-functional spaces like hotels.

With a good grasp of the design concept and the thinking that goes with it, the rest of the disciplines including architecture, structure, mechanics and lighting all have to follow suit. At EAST, the design of the main lobby best illustrates this approach. We wanted the lobby to be dominated by two elements: the reception desk and the staircase, which represents two contrasting characters of Hong Kong, its yin and yang. We hoped the reception desk would reflect an Oriental sense of calmness. At the same time, we felt the staircase should express the city's energy and dynamism, with a labyrinth-like sculpture of steel and wood supporting a glass staircase. Our client supported this concept and commission a series of scaled models and mock-ups to study every detail to make sure it all came together.

We managed to persist with a clear sense of thinking, and overcame many difficulties throughout the work period. At a critical point, our project manager left CL3, but fortunately the rest of the team was able to step up to the challenge, and at the end the project was successfully completed. EAST was the first hotel project in which CL3 was responsible for the entire design, and we would like to thank our client Swire Group for their support and trust.

EAST HONG KONG
东隅酒店

Client : Swire Properties
Location : Hong Kong
Area : 17,000 m²
Completion Date : January 2010
Project Description : 345 Rooms Business Hotel
Photographer : Nirut Benjabanpot, Michael Weber

客　　户 : 太古集团
项目地点 : 香港
项目面积 : 17,000 m²
设计时间 : 2010年1月
说明文字 : 345客房商务酒店
摄 影 师 : 王思仰，Michael Weber

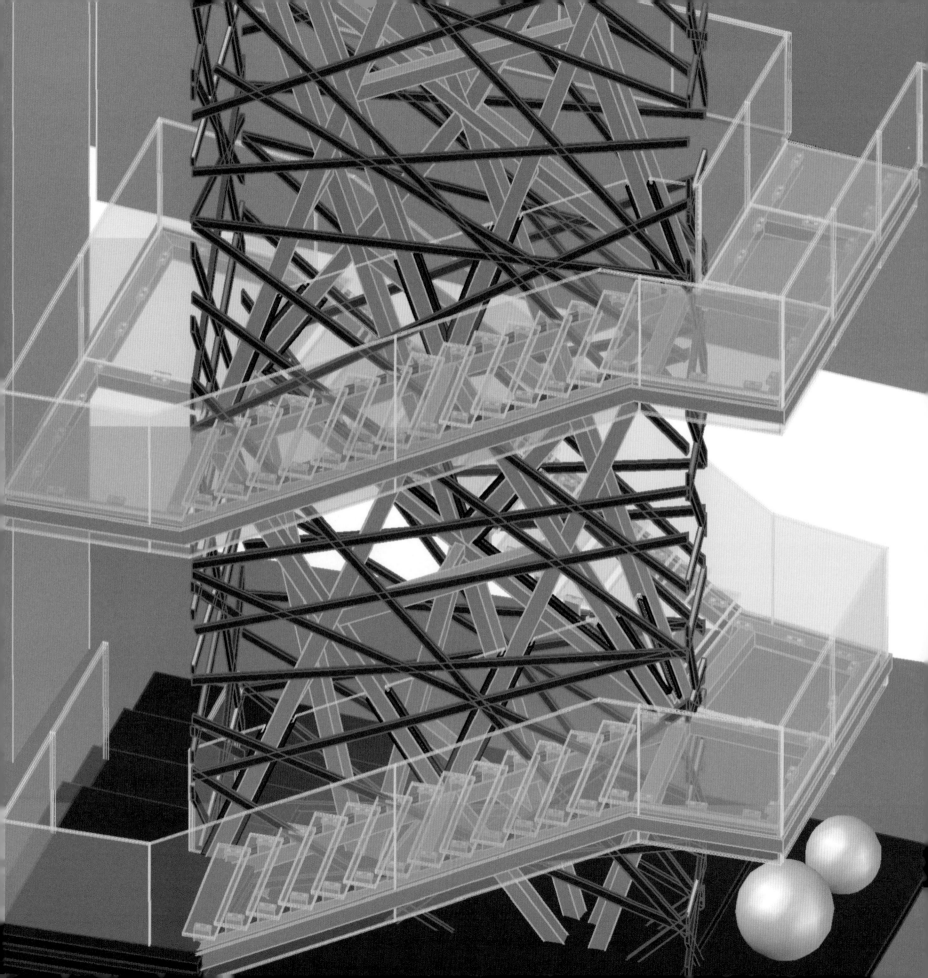

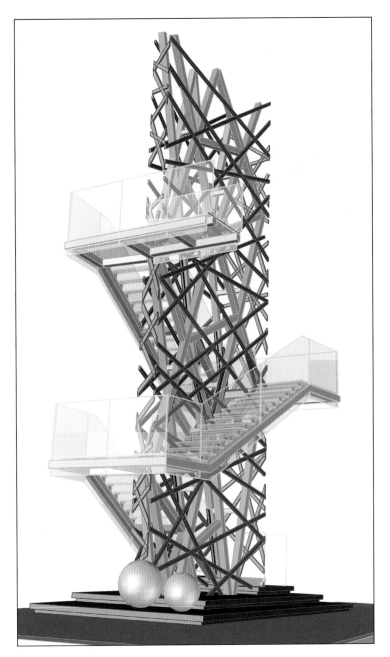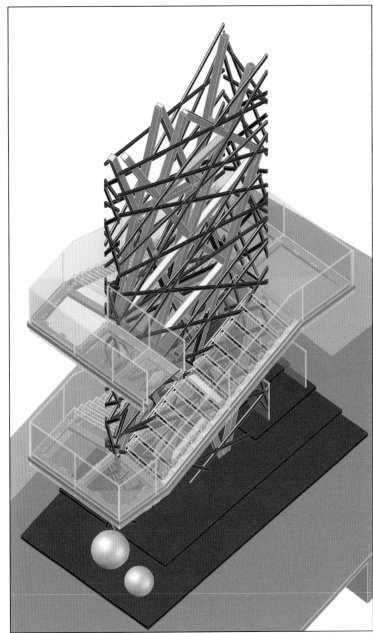

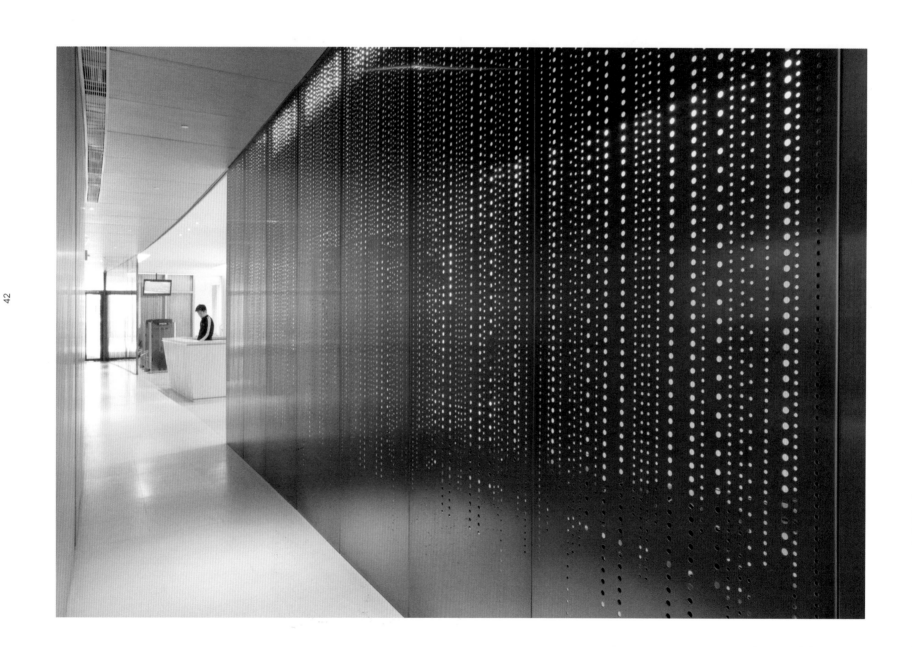

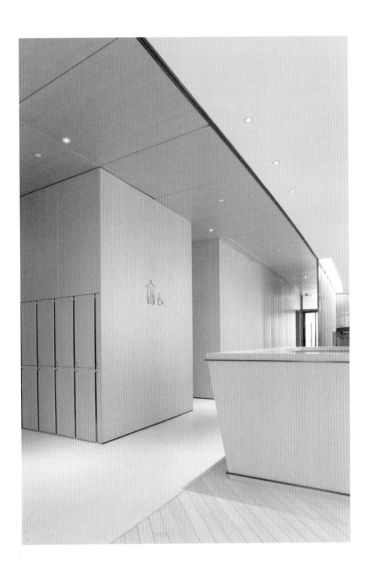

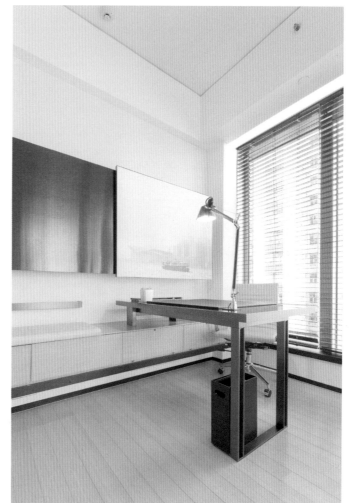

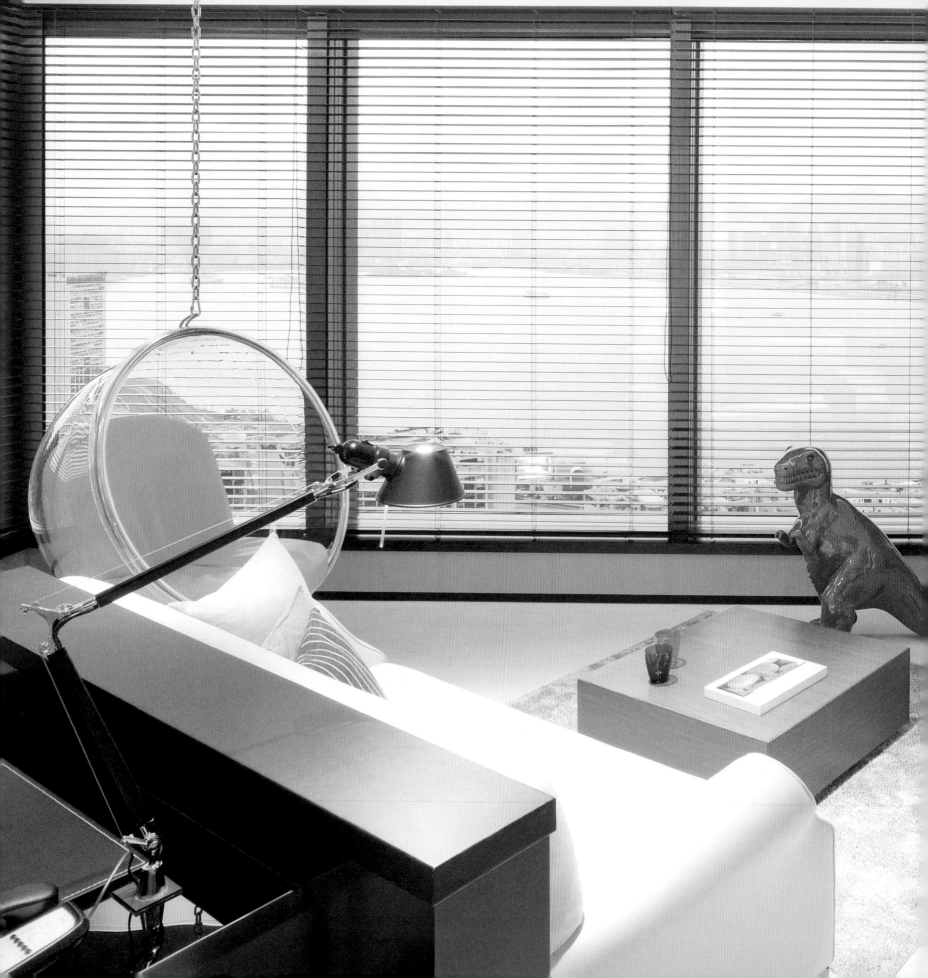

计

PLAN

计
一个庞大的计划

当客户请CL3负责新加坡滨海湾金沙项目部分设计时，项目的室内设计工作已进行了一年多。客户之前选了另外一家室内设计公司与建筑设计师合作，但两者在配合上出现问题。金沙项目美国方面的负责人很喜欢CL3在澳门为皇冠酒店设计的两间餐厅，于是希望我们能为滨海湾金沙酒店负责部分室内设计。我们考虑了很久是否接受这个设计项目，因为项目规模比我们之前做过的任何一个都要大，并且设计时间紧迫，压力很大；另一方面，如果接受这个项目，我们也要增聘员工，这会对公司的经营运作产生很大影响。我们一向的原则是不接受超出我们能力范围的项目，因为我们希望把每个项目都做到最好，而这个新加坡滨海湾金沙项目将会牵涉我们半数以上的员工。此外，项目的其他人员都是国际级顾问，只有CL3来自是香港（中国）本土。CL3深知客户的要求很高，我们是否可以应付这些工作，还是要详加考虑的。

最后，CL3决定接受这一挑战。项目规模十分庞大，是一个有近三千客房的度假酒店，属于新加坡的地标建筑。我们的设计范围包括酒店大堂及公共空间，其中有两个接待大堂、三个电梯厅、三个大堂餐饮空间，以及顶层一个天台泳池的环境布局规划及选材。我们除了要应对客户的高要求外，项目也受到新加坡政府的高度重视。

项目的开始是对整体平面进行规划，需要考虑酒店内的大量人流。整个空间长280米，高约60米，其中有两个大型玻璃中庭，如此高大的空间不可能加吊顶，所以CL3的规划除了平面上布局要合理，还要考虑布局与建筑的关系。我们在空间设计上加入了一些高差的变化，以此来分隔人流及功能，也使空间有高低交错的层次感。这种大型空间设计已跨越了单纯的室内设计，更涉及室内建筑的规划。在建筑的周边有很多水面，绿化是建筑的延伸，这部分也是我们工作的范围。我们把一些室内环境的考虑如绿化、水景等延伸到室外，使环境内外贯通，室外景观成为室内设计的一部分。

滨海湾金沙的主设计师是波士顿的著名建筑师Moshe Safdie。设计的主要汇报都是在波士顿，那里是林伟而曾工作和生活了六年的地方，所以他很高兴再回到那里。CL3第一次汇报是在2008年1月份，当时我们先空运了大量的汇报材料过去，然后再提前两天到达，以准备好汇报材料。汇报当天，雪下得特别

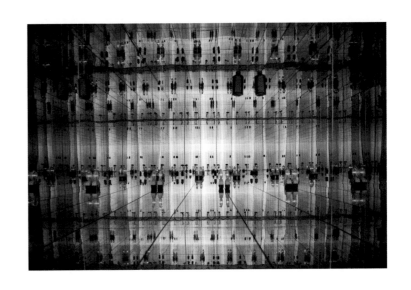

大。当天来汇报的还包括负责酒店其他项目的团队,到下午才轮到我们汇报酒店室内公共部分的设计。汇报的时间不多,幸而客户很支持我们的理念,方案很顺利地通过了。

对于这个大项目的第一次重要汇报,林伟而事后曾表示,当他汇报完成后走到外面,看到白蒙蒙一片的雪景,顿时觉得放松了下来。在汇报的地方旁边,他看到一座四层高的建筑,竟然是他二十多年前在波士顿工作时负责设计的,如今还屹立在那里。那是一座银灰色铝板的理查德·迈耶式建筑,与林伟而现在的设计有所不同,但简洁的风格却是他现在还保持着的设计理念。

像滨海湾金沙这样大型的项目,"计"十分重要,"计"不仅指是设计的计,还包括项目整体的计划:如何为设计人员安排好工作,满足项目的每一个需要;怎样利用员工不同的强项,分配到项目不同的需求上。而我们也需要操控计划的每一个部分,哪些部分需要坚持,哪些决定不能有错,哪些时间一定要把握好,每一个环节都能协助项目达至成功,以及赢取客户对公司的信任。

在这个项目中,林伟而做的一个很重要的决定,是石材的选择。这个项目单是用在地面的石材已达四千多平方米,而整个大堂的地面只用了两种石材。因为都是在连贯而整体的区域中,所以看起来需要很自然均匀,但是石材的供货期非常紧张,最后只能在几种现货石材上选定。开始时与建筑设计师的磋商是用两种对比较大的石材配合,但林伟而到石材厂看过实物后觉得并不理想,反而用两种颜色相似的石材配搭,效果会更好。当时已经没时间去争取所有人的一致通过,只能凭个人的感觉和经验作决定,并且要在石板上签名封样。当中的压力可想而知,但他也只能硬着头皮去做,而最后的效果还是得到了各方的肯定。

滨海湾金沙项目对CL3是一个巨大的挑战,它的圆满完成也让我们有了质的飞跃,使CL3可以与一些跨国大型顾问公司并驾齐驱。这个项目的成功有赖于我们的一班得力团队,及一个周全的项目计划。

Plan
A great project starts with good planning

When CL3 were asked to work on the Marina Bay Sands project, the design work had already been going on for over a year. The Las Vegas Sands management team saw our work for two restaurants at Crown Hotel in Macau (now renamed Altira Macau), so they approached us for the Singapore project. We deliberated for a long time whether to take on this job because of its massive scale, far larger than any project we had previously worked on. Also the schedule was tight, and we knew the pressure would be immense. Also, taking on this project would mean we had to employ additional staff, which would have far-reaching implications for the operation of CL3. We estimated that this project would involve over half of our staff. We had always operated under the principle to never to take on more than our existing staff could handle. The rest of the project team consisted of internationally renowned consultants. We could sense the pressure even before we had started.

In the end, we decided to accept the challenge. The project was huge in scale, consisting of a nearly 3000-room resort hotel housed within one of Singapore's iconic buildings. CL3 would work under the project architect Moshe Safdie on designing the lobby of the hotel as well as all the public areas, including two reception areas, three lift lobbies, three food and beverage outlets, as well as the furniture, fixtures and equipment for the swimming pool and surrounding areas located on the rooftop. The project is a flagship not only for the client, but also a visible icon for the Singapore government.

Our work began with planning the general layout of the floorplate, taking into account the enormous pedestrian circulation through the resort hotel. The site measured 280 meters in length, with a height of 60 meters, including two glass atriums. For such a massive space, a ceiling would not be possible, so the layout had to be planned in a rational manner, taking into account the relationship between the interior functions and the architecture. We used differences in elevation to separate human traffic flow from the programmed spaces, giving the F&B space its deserved privacy. We extended the interior design of the landscaping and pool area to unify the interior and exterior spaces, allowing the spectacular view to become part of the interior design.

The chief architect for the MBS project was the famed Moshe Safdie, who was based in Boston. This was a city that William had lived and worked in for six years, so he was glad to be back when we had to do our project presentation. He remember the first presentation in January 2008. Before leaving for the U.S., we had sent over a great deal of material by air, and arrived in Boston two days before to prepare. On the day of the presentation, it was snowing heavily. In addition to us, there were other project teams responsible for different parts of the hotel complex that also had to do presentations, and it was mid-afternoon by the time it was our turn. We did not have much time to present, but fortunately the client fully supported our ideas, and our proposal was quickly accepted.

After presentation, William stepped outside the building into the freshly fallen snow and felt quite relaxed. As he walked along, he noticed a four-storey building adjacent to the building in which we did our presentation. He realized he had designed this building when he was working in Boston over 20 years ago, and was surprised to find it still standing. Even though he had moved on with his design exploration, this silver-grey aluminum paneled Richard Meier-style structure with its clear, rational approach is a design philosophy still strongly part of the CL3 vision.

For a large scale project like MBS, team structure was an important element in our work. CL3 had to carefully strategize for the entire project, such as how to arrange the work for the design team to ensure that all the project's requirements were satisfied, and how to take advantage of every team member's experience and strengths and direct them towards appropriate sections of the project. We had to take full control of the project, and know when to be insistent, which decisions had no margin of error, what schedules had to be adhered to, because every aspect contribute to the project's success.

The decision in choosing the stone floor paving was especially challenging. The stone flooring for this project covered 4000 square meters, and the entire lobby employs only two types of stones, one for the field and one for banding highlights. Since the flooring covered connecting areas, the stone slabs must look natural and seamless throughout. The delivery schedule for these paving stone was very tight. We could only choose from a limited range of available stones. Initially, we had agreed with the design architect to use two kinds of stones with contrasting colors, but when William got to the stone factory for the selection, he found that the stock they had did not look right together. Instead, using two types of similar-colored stones was much more aesthetically pleasing. At that point, there was no time to confer with other team members, and William had to make a snap decision based on his gut feelings and experience. The pressure he felt when he signed his name on the sample was immense, but there was no other choice. Fortunately, the end result came off nicely.

The MBS project was a great challenge for CL3; its fruitful completion put the company on a new level, and allowed us to stand shoulder-to-shoulder with multi-national consultant companies. The success of this project was due in no small part to our competent team, our careful and thorough project planning, a great client, and of course the architect Moshe Safdie's trust and the opportunity he gave us.

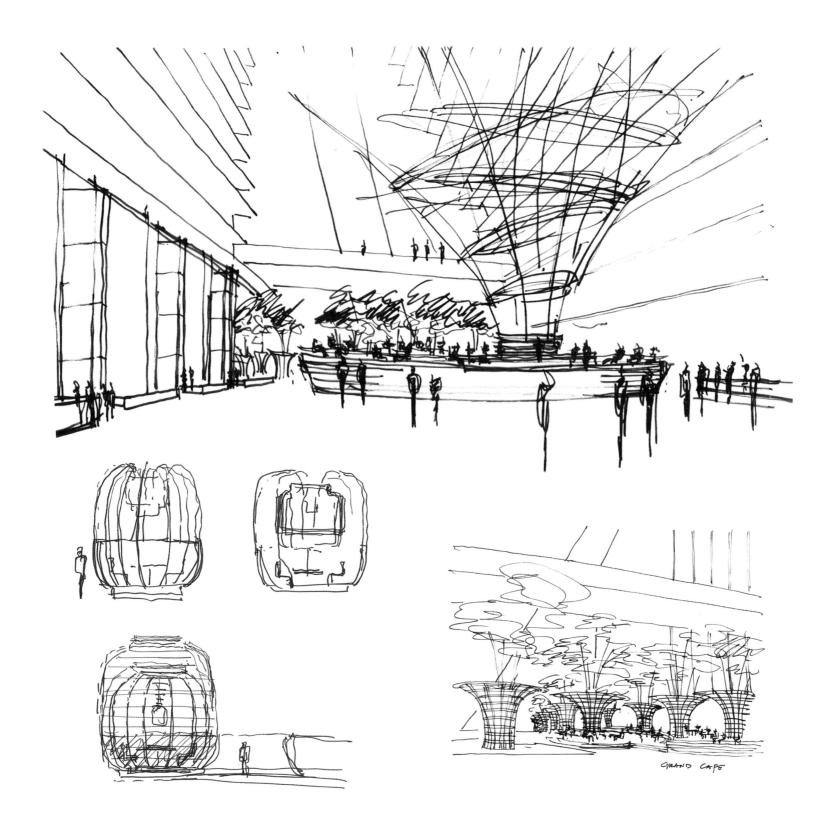

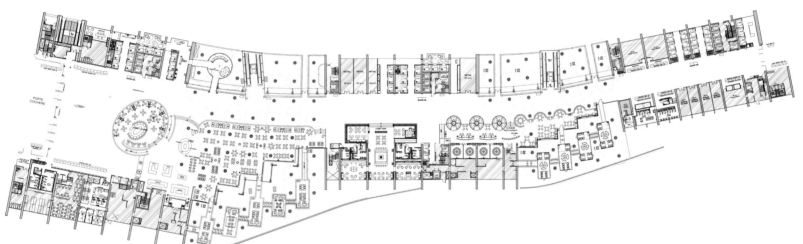

MARINA BAY SANDS
新加坡滨海湾金沙酒店

Client	:	Marina Bay Sands Pte Ltd.
Location	:	Singapore
Area	:	7,220m²
Completion Date	:	April 2010
Project Description	:	Hotel Public Area, Roof FF&E
Photographer	:	Nirut Benjabanpot

客　　户	:	新加坡滨海湾金沙
项目地点	:	新加坡
项目面积	:	7,220m²
设计时间	:	2010年4月
说明文字	:	酒店公共区域、顶层FF&E
摄 影 师	:	王思仰

展
EXTEND

展
不用常规，扩展思路

设计酒店及餐饮项目，除了功能要求之外，还要达到一定的震撼力。美国一些很成功的设计师，本身是舞台设计师或艺术家，所以能设计出很戏剧性的空间。CL3的董事总经理林伟而也经常会将他对艺术的兴趣融入设计中。其实在营造震撼力和打动人的氛围方面，CL3之前做的一些商业设计中，如NIKE的商业空间，已吸取到一定经验，特别在考虑功能需要、客人流程及兼顾营造销售气氛方面。

林伟而经常对年轻设计师说，要把握每个机会做到最好，即使最小的项目也可能会带来很大的机会。在CL3成立的初期，公司很幸运地获得了NIKE的聘用。当时CL3没有什么已完成的作品可以展示给NIKE的设计总监，但我们的一个设计师很有心思地做了精彩的汇报。那时还没有powerpoint软件，他只用了几张A3的纸板，就把CL3的架构很清晰地表达了出来，使我们介绍时十分清晰，客户也觉得我们很认真。就这样，CL3为NIKE的一间很小店铺做了设计，结果令客户很满意。后来NIKE在亚洲大规模扩大业务，他们就让CL3负责公司在亚洲一系列的专卖店及办公室设计。

NIKE对设计的要求很高，但也给设计师很大的自由度，他们很乐于尝试新的设计理念。CL3为他们设计的办公室，把很多共享空间放在办公空间内，又加入小店铺、健身、餐饮等功能，在当时十分具突破性。最后CL3凭这个设计项目拿了不少奖项，使我们在业界小有名气。

NIKE对品牌的形象十分重视，他们所做的每一步都对品牌推广有很大的设想。CL3设计他们的零售店，在功能分配、产品陈列推广、整体气氛的营造、品牌形象的展示等方面，都力图给顾客特别感受，让他们对品牌接受认可，继而刺激购买欲。林伟而将这个设计理念同时应用到了餐饮设计中。两者不论货物或食物，质量都要有吸引力，才能与空间设计相得益彰。CL3设计了很多酒店的全日餐厅，其中大部分都有开放式的自助餐，开放的自助餐台把食物与陈列品放在一起，展示出激发人食欲的美景，和精品店的货品陈列很相似。与此同时，CL3的团队中有一些拥有丰富酒店设计经验的设计师，因此他们很容易地就从商业设计延伸到了餐饮设计上。

如果说品牌店设计与餐厅设计共通，那么办公室设计就是酒店公共部分设计的缩影，两者都包含对外及对内的两方面，并且都具有接待功能及商务功能，以及有大量的人流货流要控制及分类。CL3在1996年至2006年间，为很多大型企业设计办公室，其中包括长实集团、中银集团、耐克集团、万科集团等，这些办公室都展示了其企业形象。

随着CL3在餐饮设计领域的成功，2004年香港信和集团皇家太平洋酒店委托CL3负责部分空间的室内设计。这是一个商务酒店翻新工程，CL3负责酒店的公共部分、房间改造及餐饮部分的设计。在这个项目中，我们把酒店的接待大堂与商务中心相连，让这个简单明快的空间同时担当了酒店大堂和商务大堂的角色。在酒店餐饮部分的设计中，我们打通了厨房，打造了一个很大的开放厨房区，将餐饮区改造成一个全日自助餐厅。食品的陈列方面，则加入了零售的理念，将食品陈列与空间设计合二为一；又设计了几种不同的用餐环境，包括开放、半开放及包房等，使餐厅的业务比改造前有了很大的进步。

林伟而经常使设计元素做出新的突破，并把握每个机会让客户接受他的想法。在2004年受九龙香格里拉酒店委托设计的滩万日本料理餐厅中，我们的设计加入很多艺术元素，效果十分理想。在2007年，香格里拉酒店继续委托CL3设计北京西村日本料理餐厅，这也是一个翻新工程。在这个项目中，存在很多结构上的局限，与此同时客户还希望CL3能够把控成本。我们认为餐厅大堂中间的柱子，需要装饰隐藏起来，于是采用了一种很有突破性的材料——层压板来处理这一问题。我们用了1000块层压板，建成两面1米深的墙，把柱子包在里面；在没有柱子的地方，开了一些大的洞口，通过这些洞口让空间流通，使空间更有层次感。整个设计色调很统一，几乎只有一种颜色，看上去似乎只有一种材料，很有东方的禅意，简洁之余十分震撼。西村日本料理餐厅的设计获得了很多国际大奖，使国际的杂志传媒也认识了CL3。

从酒店设计，CL3很自然地涉足水疗的范畴。2004年，上海外滩三号委托我们设计依云水疗。因为水是没有颜色的，所以我们设计了白色的空间，用水槽连通了走道，再用灯光营造出了安静的氛围。2009年，我们受委托设计了香港美丽华酒店的Mira Spa，同样使用了简洁的元素和柔和的灯光，其出色的设计为CL3赢得了很多奖项。

酒店设计日趋个性化，客户很少会像前几年一样要求空间设计得像某某酒店。每一个酒店的形象、经营手法、设计，都有可能打动客户，让他信任设计师，并带来日后的更多项目。这样，设计师就能继续发挥所长，设计出更多更突破的空间。

Extend

Extending design outside the box

Good design should not only be functional, but also aim to deliver a sense of wonder and amazement. Many successful interior designers are former theatre set designers and visual artists, therefore are able to come up with dramatic and theatrical space design but not necessarily good functional design. William, CL3's managing director has a keen interest in art. CL3's design has been attempting to extend beyond functionalism into the realm of artistic installations. In our formative years, CL3 was fortunate to have worked with large corporations like Nike, who believe that good, dramatic design goes hand in hand with good business. In designing their retail spaces, we had to satisfy functional requirements and customer traffic flows as well as creating a dramatic retail experience.

Nowadays William always tell young designers that they should make the best of every opportunity since even the smallest project can lead to great things. Soon after CL3's establishment, we received a call from Nike. At that time, we did not have many completed projects to show their design director. However, one of our trusted colleagues worked hard on our presentation. At that time there was no powerpoint, so he cleverly used a few A3-sized sheets of cardboard to present the company structure, which allowed us to make a very clear presentation convincing the client of the organization and ability. So we were given the chance to work on one of Nike's retail shops, and the results proved quite satisfactory. Later on, when Nike began expanding in Asia, they commissioned CL3 to undertake the design of its Asian retail outlets and offices.

Nike was very demanding when it came to design, but also encouraged designers to be creative. They were very willing to try out new design concepts. The office that CL3 designed for them was very innovative for its time. We incorporated many shared spaces within the office, and included various facilities such as retail space, a gym and café. In the end, CL3 won many awards with this office design, which put us on the map.

Nike was very precise with its brand image, and every step they took contributed to the promotion of the brand. In designing their retail outlets, we made sure that the functional distribution of space, product display and promotion, atmosphere and display of brand image, etc, all worked together to give customers a unique experience that inspired their appreciation of the brand, and desire to purchase. This design thinking can be equally applied to food and beverage outlets, where, similar to merchandizing, the food display becomes the showcase. We have designed many three-meal restaurants for hotels, many offering buffet dining, where open-style buffet tables display and showcase a variety of cuisines, including counters for show stations. This has a great deal in common with retail display. For this reason, and because we had many designers on our team with extensive experience in hotel design, CL3 made a smooth transition from designing retail outlets to food and beverage facilities.

If one can say that there is much in common between designing for retail and hospitality spaces, then it is equally true that office design has a lot in common with public areas in business hotels. Both encompass reception and business functions, involve space planning for receiving visitors, and include internal and external

functions. Between 1996 and 2006, CL3 undertook the office design for many large corporations, including Cheung Kong Holdings, the Bank of China, Nike and Shenzhen Vanke Real Estate Co., all of which required very strong brand communication. This led to a very special opportunity in 2004, when we had a chance to work on the interior design for Sino Group's Royal Pacific Hotel. This was a business hotel refurbishment project, and we were responsible for revamping the hotel's public areas, guest rooms and restaurant. We applied much of our experience with corporate design to this hotel. We connected the reception hall with the business center, and used this simple yet well-appointed space for both hotel reception and business lobby. For the restaurant, we opened up the kitchen and remodeled the space as a three-meal buffet restaurant with an open-plan kitchen. In terms of food display, we incorporated concepts derived from retail display and integrated the arrangement of food with interior design. In addition, we introduced different kinds of dining environments including open-plan, semi-open and private rooms. The restaurant saw a marked improvement in business as a result of the re-design.

CL3 always try to achieve new breakthroughs in our design, and collaborate with clients to realize our ideas. In 2004, we worked for the Kowloon Shangri La Hotel on the re-design of their Nadaman Japanese restaurant. In 2007, they approached CL3 again to work on the Nishimura Restaurant in Beijing, another refurbishment project. The client wanted us to pay careful attention to cost control. At the same time, the site had certain structural limitations. There were some columns in the middle which we considered unsightly, so we employed a very innovative material-engineered plywood sheets to solve this problem. We used 1000 plywood sheets to build a meter-thick wall that encased the columns, and in the spaces between the columns we created ovoid openings through which one can see other parts of the restaurant, giving the space a sense of extension and depth. The use of material was very consistent throughout, almost entirely of a single tone so that it looked as if only one material was used. The atmosphere was one of Zen-like simplicity. This restaurant design received numerous international awards, with CL3 gaining profile in many international magazines and media.

From hotel design, CL3 extended into Spa design. In 2004, we had the opportunity to work on the Evian Spa for Three on the Bund in Shanghai. Using Evian, the mineral water as the DNA, we designed a mainly white space with water running along the atrium, while soft lighting was used to create a sense of serenity and calm. In 2009, we had the opportunity again to design another spa project, the Mira Spa, located in the Mira Hotel in Kowloon, Hong Kong. Employing a few core elements and atmospheric lighting, the effect is simple but dramatic. This design also won many awards.

In recent years, hotel operators are more and more keen to look for unique design. This has become an extension of their brand philosophy. Gone are the days when operators were derivative and wanted to play it safe. Nowadays each operator has a clear vision of their brand communication through interior design. This creates great opportunity for designers to excel, to experiment, and to extend new horizons.

ROYAL PACIFIC HOTEL
皇家太平洋酒店

Client : Sino Group of Hotels
Location : Hong Kong
Area : 2,000 m²
Completion Date : April 2006
Project Description : 637 Rooms Hotel
Photographer : Eddie Siu

客　　户 : 信和集团旗下酒店
项目地点 : 香港
项目面积 : 2,000 m²
设计时间 : 2006年4月
说明文字 : 637客房酒店
摄 影 师 : 萧润昌

MIRA SPA
Mira Spa

Client	: The Mira Hong Kong
Location	: Hong Kong
Area	: 1,612 m²
Completion Date	: November 2009
Project Description	: Spa
Photographer	: Nirut Benjabanpot

客　户	: The Mira Hong Kong
项目地点	: 香港
项目面积	: 1,612 m²
设计时间	: 2009年11月
说明文字	: 水疗中心
摄 影 师	: 王思仰

SHANGRI-LA HOTEL
香格里拉酒店

Client   : Shangri-La Hotels and Resorts
Location   : Shenzhen
Area   : 3,000 m²
Completion Date   : November 2008
Project Description : Lobby, Bar, Café, Chinese Restaurant
Photographer   : Hu Wen Kit, Wong Ho Yin, Nirut Benjabanpot

客　　户　：香格里拉酒店集团
项目地点　：深圳
项目面积　：3,000 m²
设计时间　：2008年11月
说明文字　：大堂、酒吧、咖啡厅、中餐厅
摄 影 师　：胡文杰，黄浩然，王思仰

联
UNITE

联
团队合力，打造香港最佳酒店

在CL3的创立之初，有四位合伙人。其中一位的姓氏是以字母C起头，而另外三位则是以L（林伟而也是其中一个L）起头，所以命名CL3。我们不想以合伙人的姓氏做中文名，因为这样太传统，又太像一家律师事务所，所以就采用了C及L的发音，起了思联这个名字。思是思想，而联是联合起来，也表示设计不是一个人可以独力承担的，而是要靠很多人的力量联合在一起努力而成。

在CL3的团队中，主设计师处于首要地位。在新加坡滨海湾金沙那样的大型项目里，顾问团队不少于20个，每一个团队负责把握一个不同的领域，如室内、绿化、水景、灯光、厨房、标识等，每个细节都有一个专家顾问，分工细致明确，只要每个队伍都做好自己的本分，就会有一个很好的作品出来。

唯港荟酒店的设计过程长达四年之久。起初除了建筑顾问，CL3就是决策方，自由度很高，但也因为这样走了很多弯路。一般情况下，酒店项目设计里都会有客户的项目管理公司发出书面功能定位及说明，但唯港荟酒店很特殊，它不是一个连锁酒店，而是由大学经营，董事局决策，所以没有说明，我们只能在一个很自由的范围里猜测客户的想法。过分自由不一定是好事，在设计上，我们也不希望有一个是没有要求的客户，这样会使设计师无所适从，也不知用什么来作标准。

幸运的是，客户后期聘请了一个酒店经理，他是CL3之前的一个客户，很支持我们，也开始对每一个部分提出明确的要求，其他顾问也逐渐到位。顾问多了，我们的统筹地位也明显地突出了起来。我们除了要配合绿化墙、灯光、机电、标识、艺术等顾问外，也要平衡多个施工单位（差不多分了五六个单位）以及客户的要求、成本上的控制、时间上的控制、材料上的控制、土建上的配合、政府部门（如消防、建筑处等）的要求等，设计后期，又有很多不同的艺术家、设计师等参与进来。各方的配合，全赖CL3十多个得力的队友，每天好像上战场一样，有大问题就先由高级助理解决，更大的问题才由林伟而去担当。

一个成功的项目,是要靠多方面的联合努力。其中最重要的,是客户对设计师的支持。这不仅仅在设计当时,以后的运作也是一样。现在很多设计师,做设计只为拍摄一些美丽的照片,而不在意真实的设计效果。所以人们看到项目现场时,往往会吓一跳,因为实景与照片完全不同。这是一种很不诚实的做法。

唯港荟由香港理工大学管理,包涵了教育的元素,是一所很特别的酒店。因此CL3在设计上也有了一定的压力。唯港荟酒店要有启发性,要与一般酒店有所不同。香港理工大学的设计系非常出色,所以启发设计的思考也是很重要的任务。我们希望酒店的公用部分拥有灵活的生命力和变化的可能性,甚至空间的日夜功能都可以灵活变化,例如泳池、健身部分晚上可用作酒吧及发布活动。宴会厅更要灵活,既可供宴会、婚礼,也可用作音乐会、展厅等。因此,我们用功能、理性、空间艺术感强的手法去设计,尽量减少空间隔断,简化装饰性符号。房间设计也很直接,舒适、功能齐全,令客人有宾至如归的感觉。客户引入很多软件进行电子控制,使酒店的科技功能很齐全。

唯港荟的设计希望能启发学生们,以后的亚洲酒店设计就可以这样简单,这样亲切。

唯港荟酒店十分成功,开业不到三个月已经有利润。我们很感激项目的建筑师严迅奇(许李严建筑师事务有限公司)把我们引入这个项目中,更令我们感动的,是酒店经管方对我们的信任,他们连圣诞树该放在哪里,也会先询问我们的意见,有这种联结团队的精神,就会使设计能发挥最大的作用。

Unite
The power of unity in the creation of Hong Kong's best hotel

CL3 is often asked about the meaning of the name CL3. When the company was founded, there were four partners, one of whose last name began with the letter C, while the last names of the other three started with the letter L, hence CL3. In devising the Chinese name, we did not want to use the founders' last names again because that would have been too old-fashioned and sound too much like a law firm. We used the phonetics of "C" & "L" and came up with Chinese words that sounded similar. We arrived at two words which mean to think and to unite. We believe that design cannot be done in isolation, but is always a united effort that requires many people coming together to create.

Naturally, among this group of people, there should be a chief architect who takes the leading role. For example, when working on large-scale projects such as the Marina Bay Sands in Singapore, there were no fewer than 20 consultant groups, each responsible for different aspects of the project, including interior design, water features, landscaping, lighting, kitchen, signage, etc. Each small area requires its own expert consultant, and when each team carries out its own responsibility to the best of its ability, the combined result should be more than the parts.

CL3 worked on Hotel ICON for more than four years. In the beginning, we were the only consultant on board aside from the architectural consultant Rocco Design Architects Limited. It could be said that we were given a great deal of creative freedom, but at the same time, this led to much abortive works. Usually, for hotel projects, the client's project manager would issue a project brief that specifies the design requirements. However, Hotel ICON is not part of a hotel chain, but is run by a university whose decision maker is a development board. No project brief was issued, and we could only speculate as to the wishes of the committee. In fact, being given too much freedom may not be a good thing. In design, the last thing one wants is decision by committee, because design without a clear vision is like driving without a destination.

After the Chairman and General Manager came on board, everything fell into place. The General Manager was very supportive, and he came on board with a clear vision of the hotel's operation. Other consultants were also retained at this time, and as the teams of consultants were assembled, CL3's position as the chief interior planner became more evident. Not only did we have to coordinate the work by other consultants such as the green wall, lighting, mechanical and electrical, signage and art consultants, we were also responsible for organizing the various contractors, which numbered around five or six at any given time. In addition, we had to assist other consultants in addressing the client's various requests, such as cost engineering, time management, materials and construction issues, as well as complying with government regulations such as fire safety for all specified materials and building codes. By the later stages of the design process, various local artists had become involved in the artwork installations, which required a great deal of coordination. Every day CL3 had to deal with many issues, but we had a team of around a dozen very capable colleagues to carry out this arduous campaign. We would let our senior colleagues solve the more general problems, while the directors would deal with the most serious issues.

A successful project is the result of many different parties uniting and working together. Yet by far, the most important element is the client's support, not just when the project was in progress, but also afterwards. Nowadays, many designers only care about taking a few good photos after a project is done and neglect everything else, so when one visits the actual site, it may be disappointing to realize that the actual project is very different from the photos.

Hotel ICON is a very special hotel in that it is a teaching hotel run by the Hong Kong Polytechnic University. This put a certain amount of pressure on CL3's design. It needed to be inspirational. The university has a strong School of Design, so we wanted the hotel to be able to inspire future interior designers too. We had to take an unconventional approach to the design. We wanted to tell students a good hotel need not be about glitz. The hotel's public area needed to have an adaptable dynamic for future evolvement, be able to serve multiple functions, and adapt to day and nighttime functions. For example, the pool area and the gym can be used as a bar at night, or as a venue for publicity events. The banquet hall is equally flexible, capable of holding banquets, weddings, as well as concerts and exhibitions. We came up with the silver box, inspired by a black box theatre that is adaptable, flexible and can be outfitted for different events. Our design philosophy was based on functionalism and innovation with a strong sense of Hong Kong presence. The design of the guest rooms needed to be functional, well thought out, naturally comfortable with attention to details so that guests would feel at home. The design is also complemented by great technology.

CL3 hopes that the design of Hotel ICON will instill in students the idea that Asian hotel design can be based on honest, functional design rather than on superficially glamorous decoration that is common nowadays.

We are thankful that in the short duration since opening, Hotel ICON has already become an enormous commercial success. It was shortlisted as one of the three best business hotels in Asia by Wallpaper Magazine and is voted among the best Hong Kong hotels by TripAdvisor. We are grateful for the hotel's architect, Rocco Yim (of Rocco Design Architects Limited) for bringing us into the project. We are grateful for the trust our client placed in us, recently consulting us on the Christmas ornamentation. It is only with such a spirit of trust and teamwork that the best aspects of our design can be achieved.

HOTEL ICON
唯港荟酒店

Client	:	The Hong Kong Polytechnic University
Location	:	Hong Kong
Area	:	470,000 m²
Completion Date	:	March 2011
Project Description	:	325 Rooms Hotel
Photographer	:	Nirut Benjabanpot, Josiah Leung, Luke Hayes

客　　户	:	香港理工大学
项目地点	:	香港
项目面积	:	470,000 m²
设计时间	:	2011年3月
说明文字	:	325客房酒店
摄 影 师	:	王思仰，梁耀辉，Luke Hayes

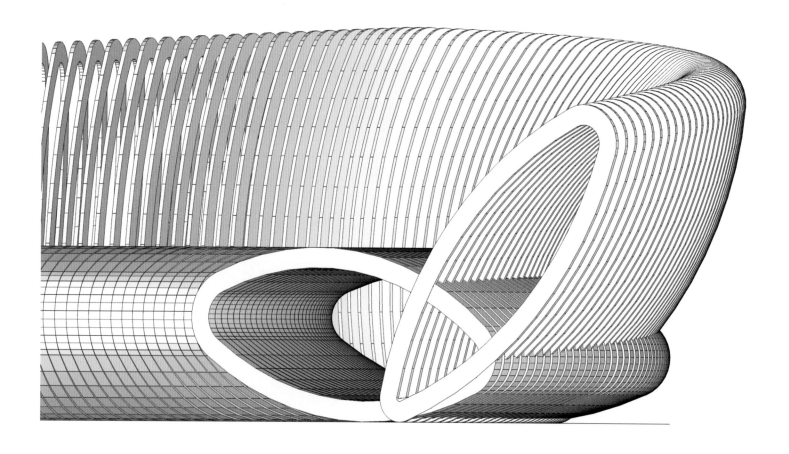

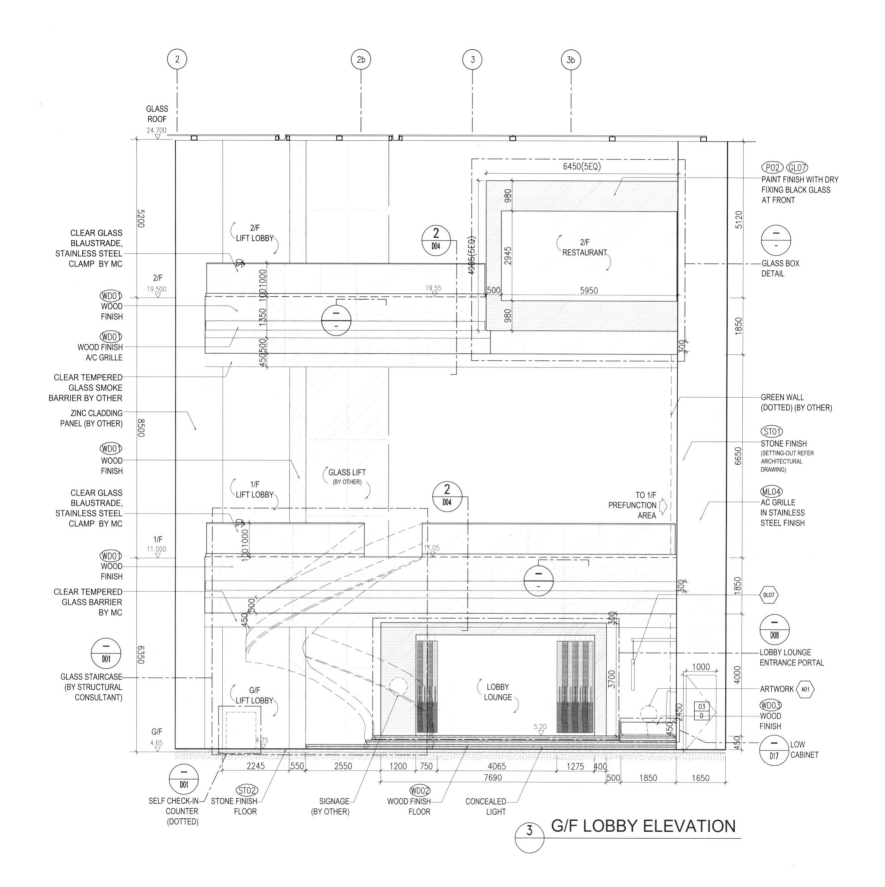

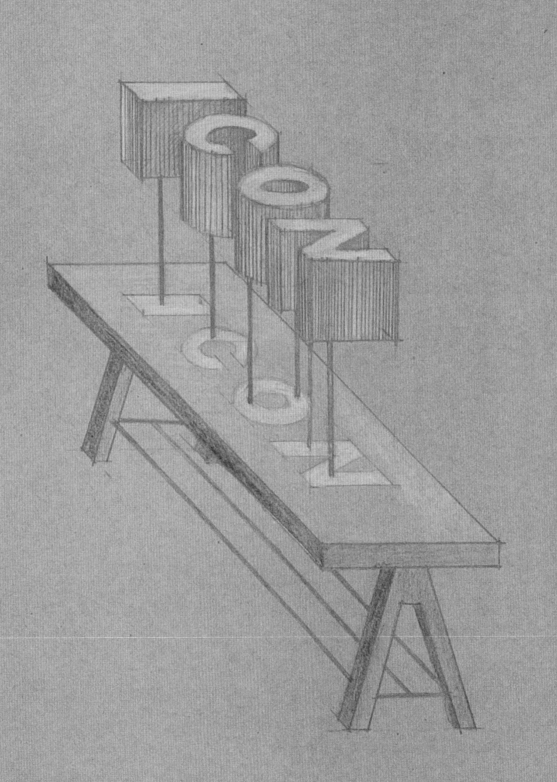

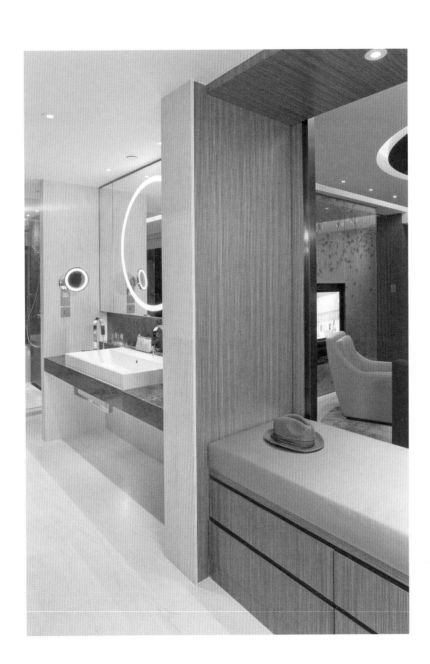

活
LIVE

活
用生活体验去思考生活环境

"形式追随功能"是勒·柯布西耶的一个很重要的理论。所以,六、七十年代的功能主义建筑师,对功能的分析很全面,甚至过于沉闷。他们研究人类生活的每一个需求,去掉没用的功能的元素。颜色的应用,也有它的功能;放置艺术品,也有它的理由。功能主义的建筑师提倡"国际风格",可以应用于任何地方,任何国家。建筑失去了文化背景,变成了国际化风格,没有了个性,也没有了文化特色。直到七十年代末,有些建筑师觉得建筑变得枯燥乏味,又开始提倡后现代主义,古典建筑的元素又回归了,甚至高层建筑也带上古典的帽子。当时美国的菲利浦·约翰逊就是其中的领导者,他从早期的极简约风格摇身一变,把所有的古典符号都放在设计里。这种建筑风格很快蔓延到全球。这个时期是建筑学最黑暗的日子。

设计与功能是不可分开的。CL3比较倾向于功能主义,但同时也认为艺术、美感、舒适感也是很重要的功能。我们会重点考虑功能上的合理性和为观赏者带来的感受,即使是艺术装置也不例外;有些不合理的设置也是故意用冲击力去制造出反理性的效果。林伟而读书时看过《建筑的复杂性与矛盾性》,书中分析了城市发展中常见的复杂及矛盾的地方。这种复杂与矛盾正好是现代社会的特征,也是生活体验的重要元素。

身在快节奏的亚洲生活,接触事物繁多,每件事都是生活的体验。因此,我们做设计时会将很多不同的生活体验也应用到设计中去。

设计酒店尤其要用自己的生活体验去构思。我们时常出差或旅游,接触酒店的机会比较多,会借此分析不同酒店的经营方式及酒店运作的合理性。例如设计酒店客房时,会站在客人的立场,切身想想他们会怎样利用这个空间。尤其在空间不大的情况下,更加要充分考虑每一个细节的做法及它的合理性。

在无锡太湖边,CL3为华润集团设计了一处位于中式建筑内的私人会所,包括餐厅、会议室、客房及娱乐空间,使用中式元素打造了一个现代舒适的空间。在这个设计中,我们在考虑每一个细节时,都以生活上的体验来把握其合理性。

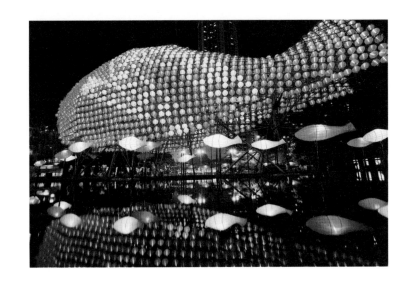

林伟而曾在美国的建筑事务所工作，对当地的公寓建筑设计有一定体会。欧美因为施工成本高，所以一般家居设计都会比较简单，反而会注重空间的运用。受了这个时期的工作经验影响，他到现在还会以功能的布局关系及空间的利用作为公寓设计的主要出发点。CL3替很多发展商设计的创意样板房，都是在这些项目的早期就已介入，经常在建筑及结构上建议开发商做出改动，使室内空间更能打动人。开发商近年来也越来越以功能布局的合理性为前提，浮夸和过于装饰的设计已不能再满足精明的客户。开发商在项目初期先请设计师做平面优化，把平面理顺了再进行室内设计，就会事半功倍。

设计创意样板房与设计酒店的共通点，在于两者都要给客户一个舒适的感觉，要满足的不是一两个客户，而是一组客户群。要让大众客户都接受设计，就需要设计师对生活有敏锐的触觉。在CL3的设计项目中，设计之初往往会首先考虑合理的功能布局，让空间的氛围大方、开放，并让参观者体会到一个舒适、温馨的生活环境，但又不失尊贵感。CL3坚持清雅、统一的设计手法，风格可以因行销方向而定，如现代、中式、东南亚或新古典，但设计的手法一定要一脉贯通整个设计。

作为室内设计师，重要的是去理解客户的需要，以及他们针对的消费者。CL3十分注重市场或经营者的需求，希望做出与生活息息相关的设计，把生活的细致融合在设计里面。功能明确合理，满足客户需求，那么设计师就可以做出更精彩的形式。

Live
Using everyday experience to reflect on living environment

Le Corbusier's most important doctrine, "Form follows function" has influenced generations of architects. Thinking that function is the driving force behind all architectural design gave rise to a very impersonal design language in the 1960s and 1970s. They looked into every aspect of human needs and left out any element that is not functional, such as emotionals. Color became coding to define functions, and placement of artwork also needs to satisfy functional need, the "International Style" was born. Architecture became an ideology, which could be applied to any place, any country. New construction technology helped shape "International Style" which did away with cultural context to allow architecture to become international, thus lacking in character and cultural specificity. In the late 1970s, many architects became discontented with "International Style", and advocated Post-Modernism. Classical elements and ornamentation adorns architecture again pediments are added to rooftops on sky-scrapers everywhere, for no reason. The American Philip Johnson was a pioneer of this movement. His style evolved from modernism in the early part of his career to one that included all sorts of classical adornments. This trend quickly spread all over the world. This period of senseless classical quotations may be one of the darkest in the history of architecture.

Design and functionalism are in-separable. CL3's design can be said to be rooted in functionalism, but at the same time we also feel that art, aesthetics and comfort, even though immeasurable, are all important functional considerations. These are essential aspects to be considered in rationalization of our design. Even art installation has its own rationality, if it is applied to a design.

The book Complexity and Contradiction by Robert Venturi, discusses the complex and contradictory aspects commonly found in urban development. In the days we live in, and the fast pace of an evolving Asian city, this aspect of complexity and contradiction is all the more relevant, and the designer's role is to resolve them in a rational way.

Living in fast-paced Asia, we encounter many different things, and each becomes part of our living experience. These experiences we have built up become the driving force for our design.

CL3's work range from hotels to clubhouses to residential show units. In each, we apply our living experience to rationalize the design. In designing for hotels, we rely very much on such experiences. We travel frequently for work and pleasure, so we spend a lot of time in hotels. We would observe the operation of different hotels, and analyze whether they are operated in a rational manner. When it comes to designing guest rooms, we would apply our experience and think about how one would utilize the space. For rooms of limited size, it is all the more important to take every detail into consideration, to come up with a great experience for the guests.

On the shore of Taihu lake in Wuxi, CL3 designed a private clubhouse for China Resources (Holdings) Co. Limited, Taihu is one of the most scenic and historical places in China. The project is housed in a traditional Chinese mansion by the lake. Our design has the look of traditional Chinese furnishings, but modernized with contemporary comfort. The clubhouse contains a restaurant, meeting rooms, entertainment rooms and guest suites. In designing each space, we employed a western point of view on functionality and comfort, based on our own living experiences. The outcome, though Chinese in style, encompasses a western approach to functionality and organization.

William used to work in an architectural firm in the United States, and has first-hand experience in the way American design firms handle residential architectural design. As construction costs in the U.S. are generally high, most homes have relatively simple interior design. Instead, designers pay more attention to space utilization. Influenced by this, CL3's design would use functional requirements and usage of space as our starting point in design. In designing numerous sales show flats for Chinese developers, we were brought onto the project at the start of architectural planning, and frequently our design requires structural alterations to the architecture to enhance the interior design spaces. As mainland Chinese consumers become more educated, developers discovered that superficial and overly-decorative designs are no longer able to attract discerning clients. Instead, design has to be well thought out functionally. Clients are more and more willing to let a well-considered plan to be the driving force of the architecture.

As interior designers, CL3 listens carefully to our client's needs. We pay a great deal of attention to our marketing team and the concerns of their business operators, and aim to design projects that would satisfy their requirements. By incorporating subtle details derived from everyday life into our work, we make the design richer and more rational. In our design we would always want to push new limits, to go beyond the basic necessities. Like life itself, we need the spice and drama. We need to strike a balance between functionality and creativity, and it is often a back-and-forth process. We always push and pull many times until a proper solution arrives, where we transcend the design to new levels.

PLATFORM (1x2)
platform (1x2)

Client	: Living Limited
Location	: Hong Kong
Area	: 530 m²
Completion Date	: March 2011
Project Description	: Gallery / Office
Photographer	: Nirut Benjabanpot

客　户	: Living Limited
项目地点	: 香港
项目面积	: 530 m²
设计时间	: 2011年3月
说明文字	: 艺术馆 / 办公室
摄 影 师	: 王思仰

RESIDENCES
住宅项目

Client : CL3's Client
Location : PRC
Project description : Private Residential
Photographer : Nirut Benjabanpot, Eddie Siu

客　户 ：私人客户
项目地点 ：中国
说明文字 ：私人住宅
摄　影　师 ：王思仰，萧润昌

建
CONSTRUCT

建
建筑由内至外

林伟而在美国完成了建筑系的学业。当时是70年代末期，还未广泛利用电脑科技。那时的学生做研究要靠到图书馆一本一本书去翻，找到有用的资料就用手记，或者影印，一般还是黑白影印，彩色影印要到后期才有。可能正是因为这点，当时的林伟而更关注的是空间、明暗对比，而不是颜色符号。当时可看到的建筑杂志也不多，因为一个建成的项目，可能要到两三年后才会登上杂志、书本等。当时一个设计元素若能刊登在美国的建筑杂志上，就代表着得到了业界的肯定。建筑是一件很认真的事，不像现在既可以在半年内建成上万平方米的建筑，也可在两年后把它拆掉。

林伟而念大学时，最受关注的建筑师莫过于勒·柯布西耶。虽然他的作品大多在五六十年代建成，但他的设计理念影响了无数日后的建筑师。就算是当今最著名的建筑师，他们的作品也满是他的影子。勒·柯布西耶是西方第一个提出建筑结构（柱、框架）可以与墙体功能分开的建筑师。东方古建筑早已是这个理论，用柱子支撑楼板，而间隔墙成为非结构墙，如此可以灵活变化地分隔功能。听来是不是很简单？这是因为几十年来很多建筑师追随这想法的结果。林伟而当时在书上看的都是他的作品，例如印度昌迪加国会大楼、一些住宅作品等。勒·柯布西耶用建筑规划来解决城市、绿化、空间及室内的问题，把以上范畴融为一体。他的理论，从人开始，由人的肢体比例关系延伸到空间、建筑、以至城市规划，这些理论在法国的马赛及印度的昌迪加得以体现。林伟而在去年初到了建筑师的朝圣之地——昌迪加，他现场看到了勒·柯布西耶的作品，才知道他的影响力之大。无论安藤忠雄的节点，斯蒂文·霍尔的空间或雷姆·库哈斯的规划，都可以在这里看到。国会大楼的震撼不止于室外——为什么建筑的外形会是这样？当走到它的内部就会完全明白。它内部功能的划分，把办公的部分放在建筑周边采光的地方，把大的国会会议中心放在中间，采用顶光，然后用走道、通廊、平台、坡道等把功能分配或相连。从建筑外面可以见到顶部凸出的国会会议中心，所有的主、次入口一目了然，主入口前的水景及它的中央延伸到远处的法院，使建筑、室内、城规、绿化成为不可分割的一个整体。这样的一个建筑师，是不会把室内或绿化交给其他顾问去处理的。

在70年代的美国学府中，康奈尔大学的建筑系是领先的代表。当时林伟而学建筑是由平面开始，首先分析地理环境对建筑所在地的影响，然后用平面解决功能上的问题，包括功能上对外、对内的需求，内部功能的布局，楼层之间的关系，主、次空间的分布，楼梯、电梯的利用。这样一步一步地把内部规划好，然后再考虑建筑周边怎样去满足内部功能，如采光的需要。最成功的设计，是从建筑的外形就可以"看"到室内功能的合理分布。所以林伟而认为，室内设计也是建筑师应做的工作。

CL3在2003年得到一个在泰国苏梅岛的项目，业主将建筑、室内及环境的设计都委托给了我们。我们用了当地的建材，并无数次到现场配合，完成了这座包括五个套房的度假别墅。在这个设计中，我们考虑了当地的气候、景观、风向来规划整体布局。屋顶根据气候的需要做了大坡顶，立面采用大量的开启窗。我们把绿化环境贯通建筑内外，把水景由前门带到泳池，视野与沙滩连接。如此一来，设计室内时已经不需要大规模改造，因为建筑及景观已经给它很好的框架，我们只需在灯光及家具、软装上用心设计即可。

这个项目的客户是外国人，她希望建筑风格简单自然，让居住的人可以放松，也有别于当地的传统泰式度假屋。我们设计的简约线条，在当时十分难以实现，每一个细节都要和承建商研究，然后在现场建造模型。客户对每一个细节都很关注，我们耗时一年多时间才完成这个项目，最终效果非常好。之后CL3在国内外接了多个现代亚洲风格的项目，包括上海九间堂的九百平方米样板间及长沙的白沙源茶馆。

CL3在2004年承接了长沙白沙源茶馆的建筑改造及室内设计。原建筑是一个中式庭院，我们在设计上既要考虑保留建筑坡顶结构的特色，又要配合空调风管的机电问题。最后，我们与机电顾问研究，决定采用地下送风系统，保留传统的坡檐屋顶，使其成为主要的设计元素。我们增加了射向屋檐的灯光，在连廊的地方加上玻璃隔断：面向中庭的是清玻璃，面向外部的是红色玻璃。我们把中庭的传统绿化改为水景，把传统的栏杆改为直条的木格栅。这样，建筑已经营造了很强的风格语言，室内部分的设计就可以相对简单了。

2010年，CL3承接了时代地产山湖海会所的设计。会所包括羽毛球室、健身、舞蹈、艺术馆等。每一个功能区域的大小不一，空间不一，采光需求不一。最后我们就每一个功能空间设计了一个不同的"盒子"，整个建筑就由这些盒子组成。因为建筑感很强，室内空间变得十分单纯。美国又又设计公司设计了一个很特别的木制模型台用作模型展示，延续了建筑设计语言。

很多人认为CL3的设计很有空间感，也许是因为我们在每个设计中都尽量去考虑建筑本身的特点，林伟而的建筑师背景让他能够更了解每个建筑的优点，也懂得去弱化它的缺点。每个设计项目都需要设计师与结构、机电顾问进行大量沟通，这可能花费很多时间，但却可以得到更好的成果。

Construct
Architecture Inside Out

William studied architecture in the United States in the late 1970s, a time when computer use was not widespread. When doing research, he had to go to the library and leaf through book after book. If he found anything useful he would take notes by hand, or use the photocopier. The photocopies were in black and white; color photocopies didn't become available until much later. It may be for this reason we were concerned with spaces and the contrast between light and dark, rather than color tones. At that time there were not many architecture magazines around. It took two to three years after the completion of a building for it to appear in a magazine, which was an important seal of approval. Architecture was a serious matter then, unlike nowadays when it only takes half a year to put up a 10,000 square meter-structure, only for it to be torn down a few years later.

When William was in university, the architect that impressed him the most was Le Corbusier. Even though most of his works had been built in the 1950s and 1960s, his design philosophy influenced countless architects who came after him. We can still see his influence in the works of the most revered architects today. Le Corbusier, the master architect, was the first in the west to propose that structure (columns and floor slabs) could be separated from wall enclosures. This has always been the case in Chinese traditional architecture. He used columns to support flooring so that dividing walls need not be structural walls, and spaces can be flexibly separated and arranged according to function. Sounds simple enough. That is only because in the past few decades, scores of architects have been following Le Corbusier's principle. William's first exposure to his works came in books, where he came across buildings like the Chandigarh Legislative Assembly and some of his residential designs. Corbusier used architectural planning to solve problems associated with urbanization, landscaping, architecture and interiors, and brought all these disciplines together to form a seamless whole. His ideas came to be realized in Marseilles in France and Chandigarh in India. William visited Chandigarh last year, a pilgrimage for many architects; when he saw Le Corbusier's work up close, he came to truly appreciate his massive influence. Whether it is Tadao Ando's detailing, Steven Holl's use of space or Rem Koolhaas's planning principles, these elements can all be found at Chandigarh. The impact of the Legislative Assembly is not just the exterior: why is it built this way? When one steps inside it all becomes clear. The offices are placed on the sides of the building to take advantage of the daylight, while the assembly meeting hall sits in the middle, lit from the skylight above. The walkways, corridors, landings and ramps are used to divide or connect spaces according to function. From the outside one can immediately see the assembly meeting hall with its protruding roof, so that the hierarchy of the building and their functions are immediately clear. The reflecting pool at the main entrance extends the central axis of the building to the High Court building in the distance so that the architecture, interior design, city planning and landscaping come together as an integrated entity. Such an architect would not leave his interior design to other consultants.

Among all the universities in the United States, Cornell is ranked No. 1 for its undergraduate studies in architecture. When studying architecture, William was taught to start with the analysis of the relationship between the context and the architecture, then to analyze the program requirements and functional problems, including exterior and interior functions, in a three-dimensional context, the arrangement between main and secondary functions, and then to connect these functions with vertical and horizontal circulation. In this way, one gives careful thought to planning the interior of the building before considering how the exterior will complement the internal needs, which then affects the fenestration treatment. In good architecture, one can "see" the rational layout of the interior by looking at it from the outside. That is why william has always felt that interior design should be the job of architects.

In 2003, CL3 did a project in Koh Samui, Thailand, which integrated architecture, interior design and environment. We visited the site numerous times, and used local materials to construct this holiday villa with five suites. We took into consideration the local climate, scenery and wind direction in planning the overall layout of the villa. We kept in mind the local climate and created a sloped roof for the rain and installed plenty of tall windows that let in fresh air. We integrated the landscaping with the interiors, and incorporated water features that visually connect the front of the house to the infinity edge swimming pool, allowing an uninterrupted view of the beach. We did not need to do much with the interior space aside from lighting, and placement of furniture and soft furnishings, because the architecture and natural scenery already provided an excellent framework.

CL3's client, who was from England, had requested that the architecture be simple and natural so that visitors to the house would feel relaxed, while at the same time making the house stand out from other traditional Thai holiday villas in the area. At that time, the Contemporary Asian style with clean detailing that we opted for proved very difficult to achieve in Thailand. Every detail had to be ironed out with the contractor, and everything has to be mocked up on-site. The client was very detail-oriented, and we had to spend over a year on site for the project. The end result was very good, and subsequently we did many more projects in this modern Asian style in and outside of China, including a 900-square meter show house in Shanghai and the Baishayuan Tea House in Changsha.

CL3 took on the Baishayuan Tea House in Changsha in 2004, which involved the remodeling of the entire building and its interior design. The original structure was a Chinese-style courtyard mansion, and we were confronted with the problem of wanting to preserve the tiled roof structure, yet at the same time needing to accommodate the requirements of air conditioning, mechanical and electrical functions. In the end, after conferring with the M&E consultants, we developed a system of underground ventilation which allowed us to keep the sloping tiled roof structure intact, the main design element of the tea house. We installed lighting to illuminate the beams, and enclosed the corridors with glass. On the side facing the central courtyard we used clear glass, whereas the side facing outside was enclosed with red tinted glass. Instead of trees (as is customary), a fishpond was introduced into the courtyard, while vertical wooden slats took the place of railing. The architecture already created a very strong visual framework, so the interior design could be relatively simple.

In 2010, CL3 designed a clubhouse and commercial complex for Times Property in Zhuhai, which contained many facilities such as a badminton court, fitness room, dance studio, art gallery, etc., all of them with different functional, size and lighting requirements. In the end we built various 'boxes' each suited to each function, stacked and linked by stairs, ramps and foot bridges. Because the architecture was so strong, the interior design could be clean and simple. The then US-based virtual firm openUU designed a model platform that reflected the architectural language of the building. This unique wooden platform could be used both as a display stand and seating.

Many clients feel that CL3's designs exude a special sense of spatiality. Perhaps that is because we always take into consideration the characteristics of architecture and space in our design. William's training as an architect allows him to understand the spatial quality of each structure and how to use it to enhance the interior design. CL3 may not be the favorite of structural and systems consultants, because often we make them change their design to suit ours, but then shouldn't the architect be the leader of the team after all.

TIMES CLUBHOUSE & COMMERCIAL COMPLEX
时代地产商业综合会所

Client : Times Property
Location : Zhuhai
Area : 20,000 m²
Completion Date : November 2010
Project Description : Clubhouse and Commercial Complex
Photographer : Nirut Benjabanpot

客　　户　　：时代集团
项目地点　　：珠海
项目面积　　：20,000 m²
设计时间　　：2010年11月
说明文字　　：会所及商业综合大楼
摄　影　师　：王思仰

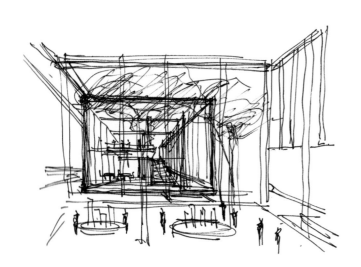
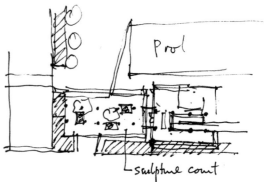

sculpture court

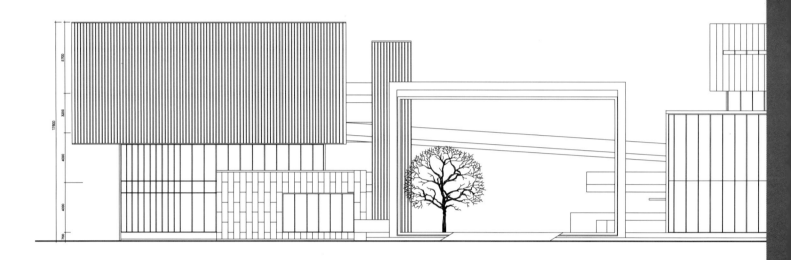
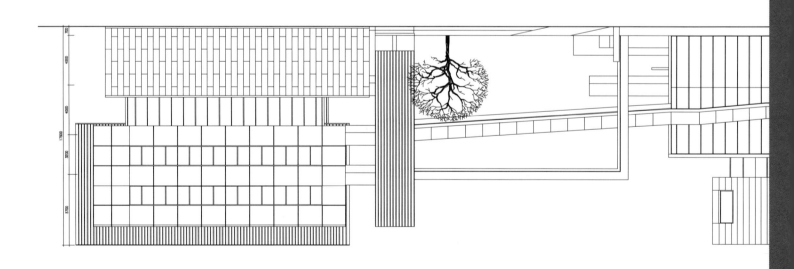

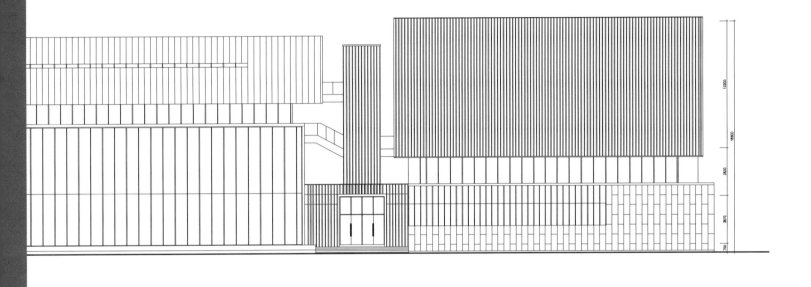
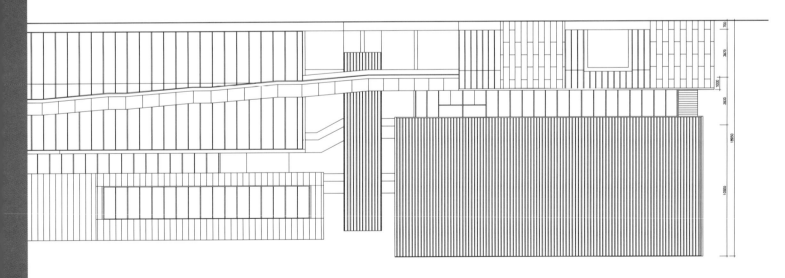

GREENLAND SALES OFFICE
绿地北京大兴售楼处

Client : Greenland Group
Location : Beijing
Area : 1,230 m²
Completion Date : 2011
Project Description : Real Estate Showroom
Photographer : Nirut Benjabanpot

客　　户 ： 绿地集团
项目地点 ： 北京
项目面积 ： 1,230 m²
设计时间 ： 2011年
说明文字 ： 房地产展示厅
摄 影 师 ： 王思仰

YANLORD LANDMARK
仁恒置地广场

Client	:	Yanlord Land (Chengdu) Co., Ltd
Location	:	Chengdu
Area	:	212,000 m²
Completion Date	:	2010
Project Description	:	Commercial Retail Complex
Photographer	:	Nirut Benjabanpot

客　　户	:	仁恒置地（成都）有限公司
项目地点	:	成都
项目面积	:	212,000 m²
设计时间	:	2010年
说明文字	:	商业及零售综合大楼
摄 影 师	:	王思仰

24-39

TWIN VILLAS
泰国度假小屋

Client   : Tropical Resort Development Ltd
Location   : Koh Samui, Thailand
Area   : 7,000 m²
Completion Date : December 2004
Project Description : Private Residence
Photographer   : Eddie Siu

客　户   : Tropical Resort Development Ltd
项目地点 : 泰国苏梅岛
项目面积 : 7,000 m²
设计时间 : 2004年12月
说明文字 : 私人住宅
摄影师   : 萧润昌

BAN SURIYA HOUSE
泰国度假别墅

Client	:	Confidential
Location	:	Koh Samui, Thailand
Area	:	220 m²
Completion Date	:	August 2003
Project Description	:	Private Retreat House
Photographer	:	Eddie Siu

客　户	:	保密
项目地点	:	泰国苏梅岛
项目面积	:	220 m²
设计时间	:	2003年8月
说明文字	:	私人住宅
摄影师	:	萧润昌

东
EAST

东
设计启自内心文化

当林伟而在美国念书的时候，大学论文与研究生论文都是以中国建筑及规划为题的。可能因为身处西方，所以当时的学生反而对东方文化产生浓厚的兴趣。他当时的导师们，也十分热爱东方文化，所以给了他很大的支持及指导。在那时，林伟而发现很多外国的著名建筑大师，像弗兰克·劳埃德·赖特、卡罗·斯卡帕及密斯·凡·德·罗等，都深受东方建筑的影响。

东方建筑与西方建筑的建构逻辑存在一定的差别。西方建筑更加注重建筑本身，试图令它高大、宏伟；为了达到巧夺天工的效果，再用装饰掩饰结构。哥特时期的建筑，就是典型的案例。东方建筑则更倾向于诚实的表达方式，因此柱、梁、屋檐等都是外露的，房间的外墙都是窗，把外面的景观引入室内。这一点在江南的房子表达得尤其显著。相对西方建筑，东方建筑不会过于注重装饰，或者说建筑就是装饰：一幅画或一幅九龙壁，都是一个陈列品，不是特意设置其中以掩饰建筑本身的特征。

美国建筑师弗兰克·劳埃德·赖特受到东方建筑的启发，设计了一些具有向外延伸的大坡顶的建筑，这在很大程度上启发了大学时期的林伟而。

东方建筑非常注重室内、外的关系，如四合院建筑的房间与庭院，两者是不能分割的整体。室外空间也是生活空间的一部分，如江南建筑会把景观"借"入室内。因为这种关系，东方建筑与环境有着很密切的联系。一个山居隐士，一定会就地取材，用毛竹建造房子；而西方的民居则会用砖来建造，这两者之间有着很大的区别。

林伟而希望寻找的东方设计语言，不是符号上的语言，而是看不见的精神上的语言，如空间中怎样带来一种大气和宁静的氛围。例如自然材料的利用、对称平衡、阴阳对比、室内外的关系、借景等，都是富于东方哲理的设计思想。

在2006年，林伟而参加了威尼斯建筑双年展，给了他一个探讨东方建筑语言的好机会。他设计了一件用竹子作主要材料的建筑，并带了五名香港的搭棚技师到当地搭建这个装置。在竹棚内，林伟而加了30只霓虹光管代表香港的霓虹灯，及中国农历每月的30天。这件艺术装置探讨了自然（竹）与科技（霓虹灯）的关系，这种阴阳的对比构建在一个很有规律的框架中，令作品既有东方文化的韵味，又很有现代气息。这件作品使得林伟而展开了一连串的艺术装置之旅，包括2009年在荷兰的"游亭"，也是一件探讨东方设计语言在西式建筑内运用的装置。

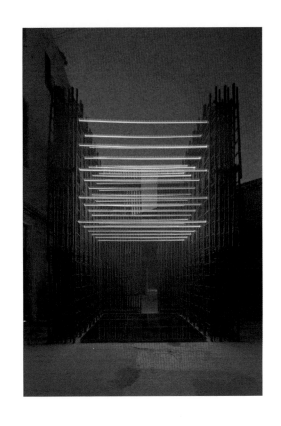

除了建筑装置外,CL3也将东方的建筑语言运用在一些小型的建筑项目中。其中成都万科的"金域西岭"项目,是一个现代中式的售楼处兼会所。在这个项目中,我们取材中式四合院的建筑模式与绿化的交错,用很统一的语言贯通了内外环境,采用红色纤维板做成建筑外墙。成都的天气适合绿色植物的生长,红墙绿叶,再加上水面的倒影,构成了一个很强的东方建筑环境。

2003年,CL3为九龙香格里拉酒店设计滩万日本餐厅。我们用沙在墙上做出波纹,加上了灯光,效果非常好,表达了浓郁的东方式空间的氛围。之后我们又做了其他的日式餐厅、中式餐厅、以及东南亚式的度假屋及水疗的设计,这让我们深入了解了东方的设计理念。在材料的运用上,我们经常选用一些贴近自然的材质,用金、木、水、火、土等元素突出材质的特性,以简约的线条营造空间感,令灯光与材质的配合烘托气氛,并配置艺术品来代表当地文化。

身在亚洲,令CL3有很多机会做东方元素的设计作品。我们认为,首先应当了解东方人的功能需求,其次则是怎样适当地突出东方的设计理念。在设计日式餐厅时,我们会先研究了解日本的饮食文化及日式餐厅的运作程式,从而避免断章取义,做得好看却不好用。同样的,设计中餐厅也有它的一套功能要求,甚至设计国内、香港或澳门的中餐厅也会随着不同区域顾客的要求而各有区别。我们为澳门皇冠酒店设计中餐厅"帝影楼"时,客户需要很多豪华包房,而大厅的设计也要有足够的空间去营造一个高级、私密的环境。于是在这个设计中,我们加入了一些传统的符号,如云海的图案、织锦的灯笼,又以中国古代花鸟画作为重点艺术元素,设计了一幅16.3米长、用水晶珠做成的立体屏风。我们与艺术顾问就此探讨了很久才达到要求的效果,于是这个屏风几乎成为了我们所有设计中最具象征意义的作品。

近年来,全球都在向东看。现在东方的强劲经济环境,提供了很多设计的好机会,也可以使我们做出具有亚洲特色的作品。CL3十分高兴有这些机会,去寻找这"东"的精神,而不仅仅是"东"的符号。

East
Design starts with one's own culture

When William Lim, the founder of CL3, was studying in the United States in the late 70's, he nurtured a keen interest in Chinese culture. William did both his undergraduate and Master's theses on Chinese architecture. At the time, he studied with a group of Cornell University professors who had done considerable research in this area. Frank Lloyd Wright, Carlo Scarpa and Mies van der Rohe, whose works are all influenced by the architecture from the East, greatly influenced him.

Eastern and Western architecture have certain basic differences. Whereas Western architecture, since the Renaissance and Gothic period, has focused on building upwards to reach the sky, often dematerializing the structure to give it the lightness to defy gravity, Chinese architecture tends to be more expressive of its relationship with its surroundings and how buildings are rooted to the ground. The architecture often deals with a sense of yin and yang, with the structure being one and nature the other. Structural elements like columns, beams and eaves are exposed. The enclosures and windows are independent from the columns structures and act as weather protection and for spatial organization. Relative to the West, Chinese architecture lacks ornamentation, or, more precisely, the ornamentation is in elements like carvings, paintings, and screens that are expressed independently from the structure itself.

The American architect Frank Lloyd Wright was greatly influenced by Japanese architecture. He employed large sloping roofs that extend outwards to engage his architecture with the landscape. His Usonian houses utilized thin, tracery-like windows, not unlike the Suzhou garden houses, to frame the outdoor views for the viewers inside.

Eastern architecture puts great emphasis on the relationship between indoors and outdoors. For example, in traditional Chinese courtyard houses, the interior spaces and the open courtyards are inter-related and inter-dependent. The living spaces often extend from the interior to the courtyards. An important part of the Suzhou garden concept is "borrowed" views. By framing particular landscape features with a specially shaped window or door openings, the exterior landscape is "borrowed" or extended to the interior space. Consequently Chinese architecture strives for a connection between the built environment and nature.

In CL3's work, we hope to discover an Eastern architectural vocabulary; we look for not literal interpretation, but a spiritual language that goes beyond what is visible. For instance, ways of creating spaces that conjure elegance and tranquility as well as the usage of natural materials. In addition, the harmony and balance of yin and yang, the relationship between indoors and outdoors, the framing and borrowing of exterior scenery are all part of the philosophy and design thinking we often employ in our work.

In 2006, William took part in the Venice Architectural Biennale for the first time, which gave him an excellent opportunity to explore Eastern architectural language. He designed a structure based on Hong Kong's bamboo scaffolding technique, and then brought five bamboo scaffolding workmen from Hong Kong to Venice to assemble it. Within the bamboo scaffolding, there were 30 neon tubes to represent Hong Kong's neon lights as well as the 30 days of the lunar calendar. This installation explored the relationship between traditional bamboo scaffolding and its contemporary interpretation, and the contrast between yin and yang, all of which are set within an ordered structure, thus giving the work a contemporary Eastern aesthetics. In 2009, William exhibited a work in the Netherlands entitled "Drift Pavilion", an installation made of paper envelops, which explored Eastern design language within Western architecture. With these works, William and CL3 began to include the concept of art installations in our projects.

In addition to installations, CL3 also applied Eastern aesthetics in some small-scale architectural projects, including the sales office-cum-clubhouse of a residential development for Vanke Group in Chengdu. We used a combination of Chinese courtyard house planning concepts and Chinese landscaping to create an environment that integrated indoors and outdoors. We used a red cement plaster board with a Chinese screen pattern as the exterior wall, and because Chengdu's climate is well-suited to plant growth, the contrast between the greenery and red walls, as well as their reflection in the water, created a strong Oriental ambiance.

In 2003, CL3 worked on the Kowloon Shangri-La's Nadaman Japanese restaurant. We used a sand texture on the walls which, together with the lighting, resulted in a very pleasant effect that also felt uniquely Oriental. The success of this project led us later on to design a number of Japanese and Chinese restaurants as well as some Southeast Asian villas and spas. For these projects, we repeatedly made use of the concept of the five cardinal elements – metal, wood, water, fire and earth to emphasize a relation of the built environment to nature. We employed simple planes and lines to conjure a sense of space, and allowed the play of light and texture to create an Eastern touch.

Living in the East, we naturally come across many opportunities to work on Asian projects. In these projects, we would consider the aesthetic and functional requirements as they relate to an Asian lifestyle. For example, in designing Japanese restaurants, we have to have a thorough understanding of Japanese food culture and restaurant operations. Otherwise, we run the risk of designing what may look good but lacks the cultural content. Similarly, when designing Chinese restaurants we need to understand the nuances of the customers using them, which vary even between mainland China, Hong Kong and Macau. In mainland China, there is great demand for private rooms, with their own pantry and toilets. In the Ying Restaurant in Macau's Altera Hotel, the client requested that we include many grand private rooms, along with a spacious open dining hall with semi-private pockets. We added some traditional Chinese crafts such as wood carving and embroidered lanterns. A 16-meter-long by 3-meter-tall crystal screen inspired by a traditional painting becomes its focus and separated the public and private spaces.

In recent years, there is great global attention on the East. The buoyant economy in China provides many great opportunities for CL3 to extend out Eastern thinking in creativity and develop works that truly represent an Asian identity. CL3 is grateful for these opportunities that allow us to push forward with a design direction that is current, relevant, and Eastern.

EVIAN SPA
上海依云水疗

Client	:	Three On The Bund
Location	:	Shanghai
Area	:	745 m²
Completion Date	:	March 2004
Project Description	:	Spa
Photographer	:	Steve Mok

客　　户	:	上海外滩三号
项目地点	:	上海
项目面积	:	745 m²
设计时间	:	2004年3月
说明文字	:	水疗中心
摄 影 师	:	莫尚勤

NISHIMURA
西村日本料理

Client	:	Shangri-La Hotels and Resorts
Location	:	Beijing
Area	:	498.6 m²
Completion Date	:	April 2007
Project Description	:	112 Seats Japanese Restaurant
Photographer	:	Eddie Siu

客　　户	:	香格里拉酒店集团
项目地点	:	北京
项目面积	:	498.6 m²
设计时间	:	2007年4月
说明文字	:	112座位日式餐厅
摄 影 师	:	萧润昌

YING
帝影楼

Client   : Altira Macau
Location   : Macau
Area   : 751 m²
Completion Date   : May 2007
Project Description : 204 Seats Chinese Restaurant
Photographer   : Eddie Siu

客　　户   ：澳门新濠锋酒店
项目地点   ：澳门
项目面积   ：751 m²
设计时间   ：2007年5月
说明文字   ：204座位中餐厅
摄 影 师   ：萧润昌

GUI HUA LOU
桂花楼

Client	: Shangri-La Hotels and Resorts
Location	: Pudong
Area	: 734 m²
Completion Date	: April 2005
Project Description	: 130 Seats Chinese Restaurant
Photographer	: Hu Wen Kit

客　户	: 香格里拉酒店集团
项目地点	: 浦东
项目面积	: 734 m²
设计时间	: 2005年4月
说明文字	: 130座位中餐厅
摄影师	: 胡文杰

VANKE SALES PAVILION
万科金域西岭

Client	:	Chengdu Vanke Real Estate Co. Ltd
Location	:	Chengdu
Area	:	7,943 m²
Completion Date	:	October 2007
Project Description	:	Real Estate Showroom
Photographer	:	Hu Wen Kit

客　　户	:	成都万科房地产有限公司
项目地点	:	成都
项目面积	:	7,943 m²
设计时间	:	2007年10月
说明文字	:	房地产展示厅
摄 影 师	:	胡文杰

BAISHAYUAN TEAHOUSE
白沙源茶馆

Client	: Hunan Baishayuan Cultural Development Co. Ltd
Location	: Changsha, Hunan
Area	: 700 m²
Completion Date	: September 2003
Project Description	: Chinese Tea House
Photographer	: Wong Ho Yin

客　　户	: 湖南白沙源文化发展有限公司
项目地点	: 湖南长沙
项目面积	: 700 m²
设计时间	: 2003年9月
说明文字	: 中式茶馆
摄影师	: 黄浩然

CR LAND CLUBHOUSE
华润无锡会所

Client : CR Land Group
Location : Wuxi
Area : 1,500 m²
Completion Date : 2011
Project Description : Clubhouse
Photographer : Nirut Benjabanpot

客　　户 ： 上海优高雅建筑装饰有限公司
项目地点 ： 无锡
项目面积 ： 1,500 m²
设计时间 ： 2011年
说明文字 ： 会所
摄 影 师 ： 王思仰

感
APPRECIATE

感
感觉, 感受, 感谢

直至今年，CL3已经成立了20周年。20年虽不算长，却也并不短，这其中CL3经历了市场的几次大起大落，也遇到过很多困难。在创业初期，接项目不容易，有段时期眼看再接不到项目的话就要歇业，我们只好硬着头皮找所有的朋友看有没有设计工程，最后还接了一个让CL3焦头烂额的项目。但还好，这个项目的首期付款使得我们可以维持下去。也有一段时期，大约在2000年左右，CL3渐受关注，当时大部分是办公室项目，是要竞标的，我们一连串竞了十多个标，每次都是第二、三名，令我们十分气馁。所以林伟而常告诉新入行者，起初一定要捱得起苦头。

正是因为客户给了CL3很多机会，我们才得以在如此艰难的条件下生存下来。除了少数客户外，大部分客户都对我们很公平：我们付出努力，客户就会付出应有的报酬。林伟而对设计的要求很高，所以在CL3工作的设计师都是对设计很认真的。我们很感谢我们的家人，一直对我们不离不弃，默默地在背后支持。

感，是感谢，是感受，所谓有感而发。CL3认为，设计师一定要从"感"做起，通过完成的项目，让客户可以感受到他的思想：如果他是很细致的，他的项目一定很细致；如果他很粗犷，项目也会很粗犷；如果他很有想法，项目中也一定可以感受得到。所以一个项目不论风格，不论造价，只要配合适合的设计师，就会带来好的设计。

通常人们去参观一所酒店，尤其是城市中的酒店，很少会注意它的外型，但一定会去评论它的室内设计。近年来，CL3的很多客户都会采取室内设计主导建筑设计的手法，或者由室内设计师先做平面布局优化，然后再完成建筑设计的部分。因此，室内设计师比建筑设计师要有更敏锐的感觉，像颜色的对比、灯光的明暗、材质的感觉、家具的舒适等，都会直接影响使用者的感受。而"感"在酒店或餐饮项目设计中尤其重要，因为它的设计就是品牌的形象，不可能开业后再去更改，所以一定要在设计期间配合好。

餐饮的设计首先要注意功能布局的要求。尤其在全日餐厅的设计中，座位的安排十分重要。在CL3设计的新加坡滨海湾金沙的全日餐厅中，包括了四百多个

座位。一个如此大型的餐厅，既要考虑客人的舒适感，也要兼顾座位的灵活性。一般酒店的餐厅是以二人及四人桌为主，而桌的组合可变成六人、八人或以上。除了灵活性的要求，我们还考虑了一些用餐环境的变化，让客人可有多种选择，例如喜欢热闹的，或喜欢清静的客人，都能找到自己喜欢的座位。在新加坡滨海湾金沙这间餐厅，因为空间很大很高，缺少了常规的舒适尺度，所以我们采用了很多绿化，并增加一些高低台阶，从而营造出亲切的氛围。最后，灯光照明也可以提升用餐环境：白天自然采光可以较随意，但晚餐就一定要营造出浪漫的气氛。

除了照明外，酒店日间和晚间的功能也会有所不同。澳门的皇冠酒店全日餐厅，早餐是自助形式，中午是半自助，晚间这是一个高档西餐厅。晚餐以供应葡萄酒来吸引食客，备有一个酒房及一个开放式酒吧，但早餐时客户不希望让顾客见到葡萄酒。这些多功能的要求，设计师要一一解决，以满足第一步功能的要求，然后才能考虑气氛的营造。CL3的设计中，这个餐厅早餐以简单明亮为主，晚餐则带有浪漫情调。在这种功能的分配里，气氛的营造十分依赖灯光设计带来的不同感觉。

在北京香港赛马会会所的中餐厅——北京凯旋餐厅的设计中，客户要求营造出一种轻松但又高档的氛围。中餐厅既有一个面档，也有高级包房。客人可以一家老少轻轻松松来吃碗面，也有客人用高级的包间宴请宾客，因此设计既要顾及两种需要，又要风格统一。我们选择现代的元素和跳跃的色调作为主要设计元素，用牡丹花来做出餐厅的符号。在入口处，设置了一组金属牡丹花雕塑，大厅内则有牡丹花图案在黄色玻璃上；而到了包间，空间的颜色变得更加庄重。尽管各个空间各具特色，但是统一的设计语言贯穿了整个餐厅。

建筑是实质的，空间却是无形的。颜色、材料是没有生命的，但感觉是发自内心的。做空间设计的难处在于，要求设计师让无形的空间打动人心。在已经走过的这二十年里，CL3不断地探索这种感觉，希望每一个项目都有所突破，有所惊喜。我们期望未来的二十年，有更多机会去追求这种空间设计的感觉。

Appreciate
Feeling, Sensation, Appreciation

This year marks CL3's 20th anniversary. Twenty years is not so long, but long enough for us to experience lots of ups and downs. We survived a few economic downturns and challenges. In the beginning, finding clients was not easy. There was a time when CL3 was about to go out of business if we could not land some projects. We had to desperately call everyone we knew to see if they had any design work. In the end we took up a project that turned out to be a nightmare, but at least the fee we received allowed us to keep our doors open. Around the year 2000, CL3 was starting to get some recognition for our corporate design. For corporate projects it was common for clients to ask for free design pitches, and for a young firm, we had no choice but to comply. There was a time then when we made over a dozen bids in a row, but each time we came in second or third, which was very disheartening. Nowadays Willliam tells newcomers to the profession that they must be prepared to endure many hardships, especially at the beginning.

CL3 are thankful to our clients for giving us opportunities to produce good work, which enabled CL3 to remain in business to this day. Aside from a few exceptions, most clients have been fair to us: we do the work, and they pay us. William is very critical with design, so designers who work at CL3 have to be very serious about design. We would like to express gratitude to our family and friends, who have always stood by us and supported us in every way.

To appreciate is to give thanks, but also to feel and sense. CL3 think good design comes from a designer's appreciation and sensitivity to the design project. If he or she is a detail-minded person, then the project will be filled with details, and if he or she is a forceful person, the project will be filled with energy. If the designer has a lot of ideas, that would come through in his or her work. Therefore, no matter what the style or the budget of a project, it will turn out well as long as it is matched up with the right designer, and the right client.

When one visits a hotel, especially if it is located in the city, one may not pay attention to its exterior, but would certainly take notice of the interior design. In recent years, many of CL3's clients turn projects inside out by allowing the interior design to drive the architectural design, resolving first the interior functions and then letting the architecture reflect that. Therefore, we feels that the interior, especially in hotel design, is of utmost importance in creating a comfortable environment by considering space planning, appreciating color contrast, light and shade, materiality, ergonomics, etc., because each of these elements directly affects the user's appreciation of the project. The role of "appreciation" is all the more important in hotel or restaurant projects, because their interior design directly reflects the brand image. Once open for business, it would be very difficult to go back and rework a design. It is therefore extremely crucial that we resolve all issues at the design stage.

One important aspect of restaurant design is resolving the layout. For restaurants that provide all-day dining, seating arrangement is very important. For example, the restaurant at Marina Bay Sands in Singapore has over 400 seats. When designing for that restaurant, CL3 had to take into account the comfort of customers as well as seating flexibility. Most hotel restaurants consist mainly of two- and four-person tables that can be transformed into tables for six to eight people or more. Aside from flexibility, we must also create different dining experiences for people to choose from. We have to consider patrons who like to be seen and those who prefer privacy or intimacy. The architecture for Marina Bay Sands by Moshe Safdie is very grand and dominating with some very tall atrium spaces, and lots of natural light. To make a comfortable space for dining they needed to bring it down to human scale. We introduced a great deal of greenery and level changes to establish a sense of coziness. Lighting can also improve the dining environment. During the day, we need to consider shading the abundant natural light to bring out a casual resort feel, but at night, the proper ambient lighting creates the mood for a fine dining atmosphere.

Restaurants need to serve different functions at different times of the day. At the Altira Macau Hotel's all-day restaurant, Aurora, breakfast is served buffet-style, lunch is semi-buffet, while at night, the place becomes a fine dining restaurant serving Western meals with fine wines. For breakfast, the client did not want the wine display to be visible, while at night, the restaurant features a well-stocked wine cellar and open bar. The first step in the design process is to satisfy the operator's functional requirements, and with that we created the theme for the restaurant and crafted its ambience. For the buffet-style breakfast in the morning, the atmosphere is casual and bright, while in the evening, the place is transformed into a romantic and intimate restaurant. A large curtain is pulled back to reveal the wine collection.

At the Beijing Hong Kong Jockey Club's flagship restaurant, the Beijing Oi Suen, the atmosphere is casual yet elegant. It includes a casual noodle stall as well as elegant private rooms, so customers can come as a family to enjoy noodles, while others can wine and dine their guests in the private rooms. Our design needed to consider the proper ambiance for both, but at the same time strove to maintain a coherent style. Using the peony as a theme and motif, the design used contemporary clean lines with splashes of colors from the soft furnishings as the main design element. At the entrance, a bronze peony sculpture greets the guests, and the peony pattern is then repeated on yellow-tinted glass in the main dining area. In the private rooms, the color scheme is more subtle, but the design language is consistent throughout.

To create interior spaces that one can sense, feel and appreciate, one would need to go beyond the bricks and mortar of architecture, or the invisibility of space itself. Color and material are lifeless, until they are transcended by the touch of design. The magic of interior design is to transcend the materiality of bricks and mortar and give the space its soul and spirit. This is the search CL3 has undertaken in the past 20 years, and will be a pursuit that will carry us forward in the next decades.

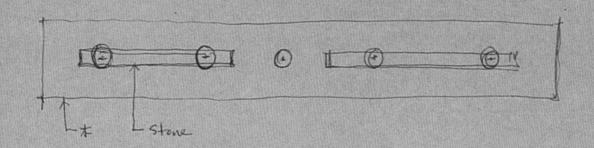

木　Stone

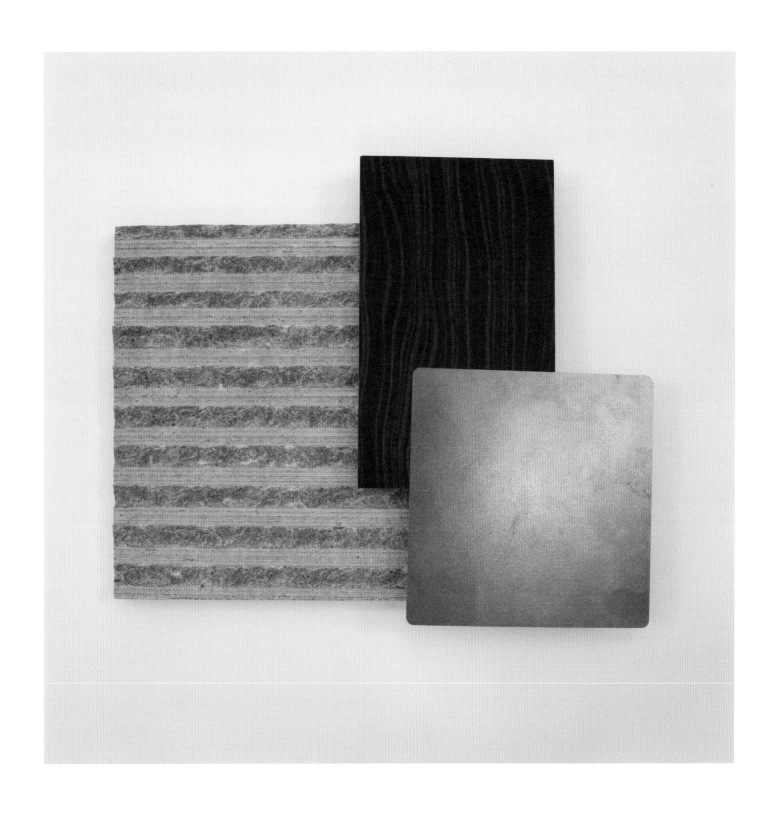

BEIJING HONG KONG JOCKEY CLUB CLUBHOUSE
北京香港马会会所

Client : The Hong Kong Jockey Club
Location : Beijing
Area : 1,287 m²
Completion Date : February 2008
Project Description : 180 Seats Chinese Restaurant
Photographer : Hu Wen Kit

客　　户 ： 香港赛马会
项目地点 ： 北京
项目面积 ： 1,287 m²
设计时间 ： 2008年2月
说明文字 ： 180座位中餐厅
摄 影 师 ： 胡文杰

NADAMAN

滩万日本料理

Client : Shangri-La Hotels and Resorts
Location : Hong Kong
Area : 457 m²
Completion Date : September 2004
Project Description : 123 Seats Japanese Restaurant
Photographer : Wong Ho Yin, Nirut Benjabanpot

客　　户 ：香格里拉酒店集团
项目地点 ：香港
项目面积 ：457 m²
设计时间 ：2004年9月
说明文字 ：123座位日本料理餐厅
摄 影 师 ：黄浩然，王思仰

AURORA
奥罗拉

Client   : Altira Macau
Location   : Macau
Area   : 813 m²
Completion Date : May 2007
Project Description : 150 Seats 3-Meal Restaurant
Photographer   : Eddie Siu

客　　户   : 澳门新濠锋酒店
项目地点   : 澳门
项目面积   : 813 m²
设计时间   : 2007年5月
说明文字   : 150座位三餐餐厅
摄 影 师   : 萧润昌

FOFO
FoFo

Client : FoFo Spanish Restaurant
Location : Hong Kong
Area : 373 m²
Completion Date : February 2010
Project Description : 88 Seats Spanish Restaurant
Photographer : Nirut Benjabanpot

客　　户 : FoFo Spanish Restaurant
项目地点 : 香港
项目面积 : 373 m²
设计时间 : 2010年2月
说明文字 : 88座位西班牙餐厅
摄影师 : 王思仰

THE RACING CLUB
竞骏会

Client : The Hong Kong Jockey Club
Location : Hong Kong
Area : 892 m²
Completion Date : May 2007
Project Description : Private Club
Photographer : Eddie Siu

客　　户 ：香港赛马会
项目地点 ：香港
项目面积 ：892 m²
设计时间 ：2007年5月
说明文字 ：私人会所
摄影师 ：萧润昌

349

ADRENALINE
Adrenaline

Client : The Hong Kong Jockey Club
Location : Hong Kong
Area : 1,700 m²
Completion Date : August 2005
Project Description : Private Club
Photographer : Wong Ho Yin

客　　户 ： 香港赛马会
项目地点 ： 香港
项目面积 ： 1,700 m²
设计时间 ： 2005年8月
说明文字 ： 私人会所
摄 影 师 ： 黄浩然

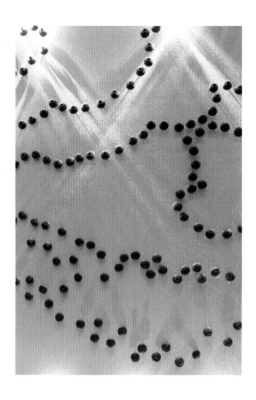

SHANG GARDEN
香乐园

Client　　　　　　: Shangri-La Hotels and Resorts
Location　　　　　: Futian, Shenzhen
Area　　　　　　　: 558 m²
Completion Date　 : January 2011
Project Description : 168 Seats Chinese Restaurant
Photographer　　 : Nirut Benjabanpot

客　　户　　: 香格里拉酒店集团
项目地点　　: 深圳福田
项目面积　　: 558 m²
设计时间　　: 2011年1月
说明文字　　: 168座位中餐厅
摄影师　　　: 王思仰

设计奖项
DESIGN AWARDS

2011

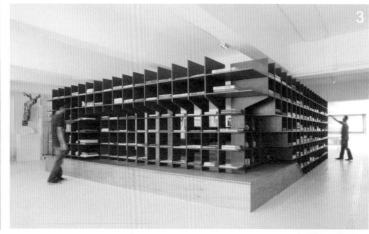

1, 2. **2011 Mid-Autumn Lantern Celebration, Lantern Wonderland Design Competition** - Gold Award Winner
中秋彩灯庆全城, 2011彩灯大观园巨型花灯比赛金奖
- Lantern Wonderland 2011 动感之娱

3, 4. **Best of Year Design Award 2011, Interior Design USA** - Residence: Urban: International
- platform (1x2)

5. **International Hotel/Motel & Restaurant Show (IH/M&RS)**,
Gold Key Awards for Excellence in Hospitality Design - Finalist (Guest Room Category)
- East Hong Kong Standard Room 东隅酒店标准客房
HA+D Awards 2011 - HA+D Award for Design Excellence Hotel Interior Design 2011
- East Hong Kong 东隅酒店

6. **Best of Year Design Award 2011, Interior Design USA** - Public Spaces/ Outdoor
- Timber Cube, Times Clubhouse & Commercial Complex 时代地产商业综合会所
The American Institute of Architects, Hong Kong Chapter - Honor Award for Architecture 2011
美国建筑师学会香港分会2011荣誉建筑设计奖
- Times Clubhouse & Commercial Complex 时代地产商业综合会所

7. **The American Institute of Architects, Hong Kong Chapter** - Merit Award for Architecture 2011
美国建筑师学会香港分会2011优秀建筑设计奖
- Hotel ICON 唯港荟酒店

2010

8. **Best of Year Design Award 2010, Interior Design USA** - Finalist (Budget Category)
   - CL3 Design Studio 思联建筑设计有限公司办公室
   The American Institute of Architects, Hong Kong Chapter - Merit Award for Interiors 2010
   美国建筑师学会香港分会2010优秀室内设计奖
   - CL3 Architects Studio 思联建筑设计有限公司办公室
   1st Annual Global Excellence Awards, International Interior Design Association (IIDA) - Winner
   (Best of Category - Corporate Space Small)
   - CL3 Studio 思联建筑设计有限公司办公室

9. **Design for Asia Award (DFA) 2010, Hong Kong Design Centre** - Merit Award
   (Culture, Public and Exhibitions - Environment Design)
   亚洲最具影响力设计大奖2010
   - Route D

10. **Design for Asia Award (DFA) 2010, Hong Kong Design Centre** - Merit Award
    (Hospitality and Leisure - Environment Design)
    亚洲最具影响力设计大奖2010
    - East Hong Kong 东隅酒店
    1st Annual Global Excellence Awards, International Interior Design Association (IIDA) - Winner
    (Best of Category - Hotels)
    - East Hong Kong 东隅酒店

11. **The American Institute of Architects, Hong Kong Chapter** - Merit Award for Interiors 2010
    美国建筑师学会香港分会2010优秀室内设计奖
    - Mira Spa

12. **The Northwest & Pacific Region of the American Institute of Architects 2010** - Citation Award
    西北太平洋区美国建筑学会优秀设计奖2010
    - Vanke Shenzhen Headoffice 深圳万科总部

13. **Asia Pacific Interior Design Biennial Awards (IAI) 2010** - Excellence Award of Club Space Design, Club Space Category
    2010亚太室内设计双年大奖赛会所空间设计优秀奖
    - Park Lane Manor – Children's Club 幸福里会所儿童俱乐部

2009

1. **Best of Year Design Award 2009, Interior Design USA** - Merit Award (Exhibition/Installation)
   - Illegal Structure 僭建
   17th Asia Pacific Interior Design Award (APIDA) 2009 - Gold Award (Exhibition Category)
   第十七届亚太区室内设计大奖2009金奖（展览类别）
   - Illegal Structure 僭建
2. **17th Asia Pacific Interior Design Award (APIDA) 2009** - Gold Award (Institution & Public Space Category)
   第十七届亚太区室内设计大奖2009金奖（学院及公共空间类别）
   - The Hennessy Public Area, Hong Kong 香港轩尼诗道256号
3. **17th Asia Pacific Interior Design Award (APIDA) 2009** - Gold Award (Office Category)
   第十七届亚太区室内设计大奖2009金奖（办公室类别）
   - K-Boxing Headquarter, Shanghai 劲霸男装上海总部
4. **The American Institute of Architects, Hong Kong Chapter** - Merit Award for Architecture Interiors 2009
   美国建筑师学会香港分会2009建筑室内设计奖
   - Vanke Shenzhen Headoffice 深圳万科总部
   17th Asia Pacific Interior Design Award (APIDA) 2009 - Silver Award (Office Category)
   第十七届亚太区室内设计大奖2009银奖（办公室类别）
   - Vanke Shenzhen Headoffice 深圳万科总部
5. **The Northwest & Pacific Region of the American Institute of Architects 2009** - Award of Merit for Architecture
   西北太平洋区美国建筑师学会优秀室内设计奖2009
   - Vanke Sales Pavilion 成都万科•金域西岭
6. **17th Asia Pacific Interior Design Award (APIDA) 2009** - Honorable Mention (Exhibition Category)
   第十七届亚太区室内设计大奖2009荣誉设计奖（展览类别）
   - Drifting Pavilion (You, Xi) 游亭（游．戏）
7. **17th Asia Pacific Interior Design Award (APIDA) 2009** - Honorable Mention (Club & Entertainment Space Category)
   第十七届亚太区室内设计大奖2009荣誉设计奖
   - CR Land Clubhouse, Chengdu 华润成都凤凰城会所

2008

  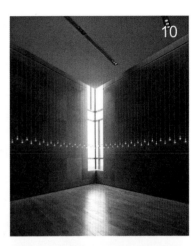

8. **Best of Year Design Award 2008, Interior Design USA** - Merit Award (Exhibition/Installation)
    - Hong Kong & Shenzhen Bi-City Biennale of Urbanism Architecture, Collective memory of a Temporary Fabric: Exploring the technique of scaffolding
    香港．深圳城市／建筑双城双年展，竹建．建筑之集体回忆 – 探索竹棚搭建技术
9. **Asia Pacific Interior Design Biennial Award 2008** - Champion (Bar Category)
    亚太室内设计双年大奖赛2008（酒吧类别）
    - Sushi Bar, HKJC 香港赛马会
10, 11. **Best of Year Design Award 2008, Interior Design USA** - Merit Award (Showroom)
    - CR Land Art Pavilion, Chengdu 成都华润凤凰城艺术中心
    16th Asia Pacific Interior Design Award (APIDA) 2008 - Gold Medal (Institution & Public Space Category)
    第十六届亚太区室内设计大奖2008金奖（学院及公共空间类别）
    - CR Land Art Pavilion, Chengdu 成都华润凤凰城艺术中心
12. **The Northwest & Pacific Region of the American Institute of Architects 2008** - Award of Merit for Interiors
    西北太平洋区美国建筑师学会优秀室内设计奖2008
    - Nishimura Restaurant 北京香格里拉酒店西村日本料理
13. **Asia Pacific Interior Design Biennial Award 2008** - Third Place (Institute Category)
    亚太室内设计双年大奖赛2008（学院类别）
    - English First Mega Center, Shanghai 英孚教育集团上海总部

2007

1. **The American Institute of Architects, Hong Kong Chapter** - People's Choice Award for Interiors 2007
   美国建筑师学会香港分会2007优秀室内设计奖
   - The Racing Club 香港赛马会
2. **Hong Kong Designers Association Award 2007** - Award of Excellence (Hospitality/Entertainment)
   香港设计师协会奖2007 – 优异奖
   - Adrenaline 香港赛马会
3. **The American Institute of Architects, Hong Kong Chapter** - Merit Award for Interiors 2007
   美国建筑师学会香港分会2007优秀室内设计奖
   - English First Mega Center, Shanghai 英孚教育集团上海总部
4. **Hong Kong Designers Association Award 2007** - Silver Award (Residential Category)
   香港设计师协会奖2007 – 银奖（住宅类别）
   - CR Land House 1 华润别墅
5. **International Hotel/Motel & Restaurant Show (IH/M&RS)** - Grand Prize Winner for Gold Key Award, Restaurants – Casual Dining
   - Nishimura Restaurant 北京香格里拉酒店西村日本料理
6. **The American Institute of Architects, Hong Kong Chapter** - Merit Award for Interiors 2007
   美国建筑师学会香港分会2007优秀室内设计奖
   - DDB Hong Kong Office DDB香港办公室
7. **Hong Kong Communication Art Centre** - Outstanding Greater China Design Awards 2007
   香港传艺中心 – 大中华杰出设计大奖2007
   - The Lantern Wonderland 彩灯大观园

2006

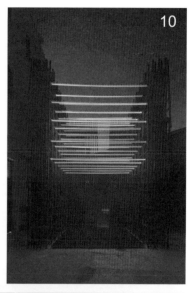

8. **Asia Pacific Interior Design Biennial Award 2006** - Winner, Corporate Category
   亚太室内设计双年大奖赛2006 – 金奖（企业组别）
   - Vanke Chengdu Commercial Complex 成都万科魅力之城销售大楼
9. **Asia Pacific Interior Design Awards 2006** - Gold Medal, Show Flat Category
   亚太区室内设计2006金奖（样板房类别）
   - CR Land House 1 华润别墅
10. **The American Institute of Architects, Hong Kong Chapter** - Honor Award for Architecture 2006
    美国建筑师学会香港分会2006建筑设计荣誉奖
    - Ladders Bamboo Installation, Venice Biennale's International Architectural Exhibition 2006
      竹梯，威尼斯建筑双年展
11. **Asia Pacific Interior Design Biennial Award 2006** - Honorable Mention, Institution Category
    亚太室内设计双年大奖赛2006 – 荣誉奖（学院组别）
    - Hong Kong Arts Centre 香港艺术中心
12. **The American Institute of Architects, Hong Kong Chapter** - Merit Award for Interiors 2006
    美国建筑师学会香港分会2006优秀室内设计奖
    - Glass Apartment 万科成都样板间
13. **Asia Pacific Interior Design Biennial Award 2006** - Winner, Hotel Category
    亚太室内设计双年大奖赛2006 – 金奖（酒店组别）
    - Royal Pacific Hotel 皇家太平洋酒店

2005

1. **Hong Kong Design Association 2005** - Silver, Exhibition/Window Display Category
   香港设计师协会奖2005 – 银奖（展览／厨窗类别）
   - Lantern Wonderland 彩灯大观园
2. **Hong Kong Design Association 2005** - Silver, Interior – Office Category
   香港设计师协会奖2005 – 银奖（办公室类别）
   - EDAW Office 易道办公室
3. **The Hong Kong Institute of Architects 50th Anniversary Annual Awards 2005** - President's Prize
   香港建筑师学会50周年纪念2005年年奖 – 会长奖状
   - Adrenaline 香港赛马会
4. **The American Institute of Architects, Hong Kong Chapter** - Honor Award for Architecture 2005
   美国建筑师学会香港分会2005建筑设计荣誉奖
   - Vanke Chengdu Commercial Complex 成都万科魅力之城销售大楼
5. **Perspective Design Recognition Awards 2005** - Honorable Mention, Best Hospitality Space (Restaurant/Lounge/Bar)
   2005透视室内设计大赏铜奖（最佳服务业设计 – 餐馆／酒廊／酒吧类别）
   - 360° Bar
6. **Asia Pacific Interior Design Awards 2005** - Silver Medal, Residential Category
   亚太区室内设计2005银奖（住宅类别）
   - Twin Villas 泰国度假小屋
7. **Asia Pacific Interior Design Awards 2005** - Honorable Medal, Restaurant/Bar Category
   亚太区室内设计2005荣誉设计奖（餐馆／酒吧类别）
   - Nadaman Japanese Restaurant, Shangri-La Island 滩万日本料理

2004

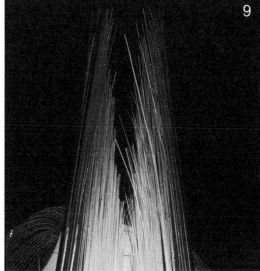

8. **The American Institute of Architects, Hong Kong Chapter** - Firm Award 2004
   美国建筑师学会香港分会2004 建筑设计贡献大奖
9. **Asia Pacific Interior Design Award 2004** - Winner, Exhibition Category
   亚太区室内设计冠军2004（展览类别）
   - Lantern Wonderland 彩灯大观园
10. **Asia Pacific Interior Design Award 2004** - Winner, Club Category
    亚太区室内设计冠军2004（会所类别）
    - Evian Spa, Shanghai 上海依云水疗
11. **Asia Pacific Interior Design Award 2004** - Honorable Mention, Showflat Category
    亚太区室内设计荣誉设计奖2004（样板房类别）
    - Show House, Shanghai 九间堂
12. **Asia Pacific Interior Design Award 2004** - Winner, Restaurant Category
    亚太区室内设计冠军2004（餐馆类别）
    - Nadaman Japanese Restaurant, Shangri-La Kowloon 滩万日本料理
13. **Hong Kong Design Centre - Design for Asia Award 2004** – Distinguished Design from China
    香港设计中心 – 优秀中国设计
    - Baishayuan Tea House 白沙源茶馆

2003

1. **The Hong Kong Institute of Architects Annual Awards 2003** - President's Prize
   香港建筑师学会2003年年奖 – 会长奖状
   - Lantern Wonderland 彩灯大观园
2. **The American Institute of Architects, Hong Kong Chapter** - Merit Award for Architecture 2003
   美国建筑师学会香港分会2003优秀室内设计奖
   - Nadaman Japanese Restaurant, Shangri-La Kowloon 滩万日本料理
3. **Asia Pacific Interior Design Award 2003** - Honourable Mention, Club Category
   亚太区室内设计荣誉奖2003
   - Purple Jade Clubhouse 紫玉山庄会所
4. **The American Institute of Architects, Hong Kong Chapter** - Honor Award for Architecture 2003
   美国建筑师学会香港分会2003建筑设计荣誉奖
   - Baishayuan Tea House 白沙源茶馆

2002

5. **The Northwest & Pacific Region of the American Institute of Architects 2002** - Award of Merit
    西北太平洋区美国建筑师学会优秀室内设计奖2002
    - Private Residence, Bangkok 曼谷住宅
6. **Hong Kong Designers Association** - Excellent Award 2002
    香港设计师协会 – 优越设计奖2002
    - Apartment, Shanghai 上海绿地集团有限公司
7. **Hong Kong Designers Association** - Bronze Award 2002
    香港设计师协会 – 设计铜奖2002
    - Furniture Series 家具系列
8, 9. **Hong Kong Designers Association** - Excellent Award 2002
    香港设计师协会 – 优越设计奖2002
    - The Marco Polo, Beijing 北京马哥孛罗酒店
    Asia Pacific Interior Design Award 2002 - Winner, Hotel Category
    亚太区室内设计冠军2002
    - The Marco Polo, Beijing 北京马哥孛罗酒店
10. **Hong Kong Designers Association** - Excellent Award 2002
    香港设计师协会 – 优越设计奖2002
    - Media Nation Office, Shanghai
11, 12. **Hong Kong Designers Association** - Excellent Award 2002
    香港设计师协会 – 优越设计奖2002
    - Shenzhen Vanke Real Estate Co. Ltd 深圳市万科房地产有限公司
    The American Institute of Architects, Hong Kong Chapter - Merit Award for Interiors 2002
    美国建筑师学会香港分会2002优秀室内设计奖
    - Shenzhen Vanke Real Estate Co. Ltd 深圳市万科房地产有限公司
13. **Hong Kong Designers Association** - Excellent Award 2002
    香港设计师协会 – 优越设计奖2002
    - Living Ltd

2001

1, 2. **The American Institute of Architects, Hong Kong Chapter** - Merit Award for Architecture 2001
美国建筑师学会香港分会2001优秀室内设计奖
- Nike Inspired I, Bangkok 曼谷耐克店铺

3. **The American Institute of Architects, Hong Kong Chapter** - Merit Award for Interiors 2001
美国建筑师学会香港分会2001优秀室内设计奖
- Top Result Promotion Ltd, Office World Trade Center 国贸中心, 通成推广有限公司

4. **The American Institute of Architects, Hong Kong Chapter** - Merit Award for Interiors 2001
美国建筑师学会香港分会2001优秀室内设计奖
- Investment Office, Shanghai 上海投资公司

5, 6. **Asia Pacific Interior Design Awards 2001** - Honorable Mention, Residential Category
亚太区室内设计2001荣誉设计奖
- Private Residence, Bangkok 曼谷住宅

7. **Asia Pacific Interior Design Award 2000** - Winner, Corporate Category
亚太区室内设计冠军2000
- Ove Arup & Partners (HK) Ltd 奥雅纳工程顾问办公室

2000

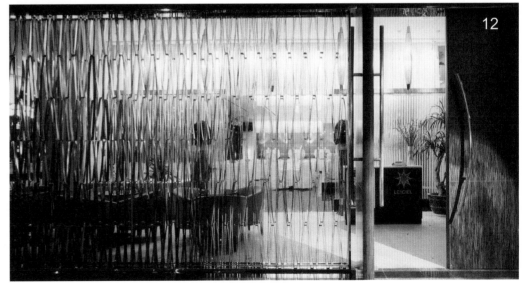

8, 9. **Hong Kong Designers Association** - Excellent Award 2000
香港设计师协会 – 优越设计奖2000
- Saatchi & Saatchi / Zenith Media Office 广州盛世长城国际广告公司

10, 11. **Hong Kong Designers Association** - Bronze Award 2000
香港设计师协会设计铜奖
- Ove Arup & Partners (HK) Ltd Office 奥雅纳工程顾问办公室

The American Institute of Architects, Hong Kong Chapter - Merit Award for Interiors 2000
美国建筑师学会香港分会2000优秀室内设计奖
- Ove Arup & Partners Office 奥雅纳工程顾问办公室

12. **Asia Pacific Interior Design Award 2000** - Winner, (Restaurant Category B)
亚太区室内设计冠军 2000
- Le Ciel Restaurant 天上人间北京法国餐厅

1998-1999

1, 2. **Hong Kong Designers Association Design Awards 1998**
香港设计师协会设计奖1998
- NIKE Office, Shanghai Mart & Bangkok 耐克公司, 上海及曼谷办公室

American Institute of Architects, Hong Kong Chapter Honor Award for Interior 1998
美国建筑师学会香港分会1998室内设计奖
- NIKE ShanghaiMart, PRC, China 上海耐克办公室

3. **The Northwest & Pacific Region of the American Institute of Architects 1999** - Award of Merit
西北太平洋区美国建筑师学会1999优秀室内设计奖
- NIKE (Suzhou) Sports Co., Inc. 耐克（苏州）体育用品有限公司

4. **Hong Kong Designers Association Design Awards 1998**
香港设计师协会设计奖1998
- Carmen's Bar & Grill, Malaysia

5. **The Chartered Society of Designers Hong Kong Design Award 1999**
特许设计师协会香港设计奖1999
- Tricon Office, Hong Kong 百胜餐饮国际集团香港办公室

6. **The Chartered Society of Designers Hong Kong Design Award 1999**
特许设计师协会香港设计奖1999
- Alliance Francaise, Hong Kong 香港法国文化协会

7. **The American Institute of Architects** - Hong Kong Chapter - Honor Award for Interiors 1999
美国建筑师学会香港分会室内设计荣誉奖1999
- Saatchi & Saatchi Office, Guangzhou, PRC. 广州盛世长城国际广告公司

8. **Hong Kong Trade Development Council Services Award 1998** - Export Marketing
香港贸易发展局服务业奖1998 – 出口市场推广

9. **Hong Kong Designers Association Design Awards 1998**
香港设计师协会设计奖1998
- HKTDC Showroom, Beijing 香港贸易发展局, 北京

1992-1997

10, 11. **Asia Pacific Interior Awards 1997** - Winner
亚太区室内设计奖1997
- Nike, Corporate Office Design 耐克办公室

12. **The Chartered Society of Designers Hong Kong Design Award 1996** - Commercial Interior Design – Section Award
特许设计师协会香港设计奖1996 - 商业室内设计
- Nike Offices, Shanghai 耐克公司上海办公室

13. **The Chartered Society of Designers - Interior Design Award 1993**
特许设计师协会香港设计奖1993
- Club Casablanca (Karaoke Disco), Beijing PRC 北京卡隆布兰卡歌舞厅卡拉OK
The Chartered Society of Designers - HK Seibu Interior Design Award 1993
特许设计师协会香港西武室内设计奖1993
- Club Casablanca (Karaoke Disco), Beijing PRC 北京卡隆布兰卡歌舞厅卡拉OK

团队
TEAM

为思联闪亮过的每一颗星
All the stars that shine for CL3

Bernadette Mary Acuna, Shamim Ahmadzadegan, Jane Arnett, Betty Au Yong, Jonathan Banner, Nirut Benjabanpot, Barnaby G Bugden, DoubleSun Cai, Joe Cai, Morag Cameron, Catherine Jane Campbell, Benjamin Chan, Caroline Chan, Celia Chan, Christine Chan, Clara Chan, Conyee Chan, Eddie Chan, Gary Chan, Johnny Chan, Liz Chan, Nicky Chan, Patsy Chan, Penny Chan, Tik Chan, Damon Chang, Aaron Chau, Pest Chen, Stephanie Chen, Lydia Cheng, Wilson Cheng, Amos Cheung, Cindy Cheung, Darcy Cheung, Dennis Cheung, Rose Cheung, Vani Cheung, Joe Chiu, Nerisa Choi, Anthony Chow, Billy Chow, Dennis Chow, Janet Chow, Kelly Chow, Kin Chow, Martin Chow, Philip Chow, Raymond Chow, Samuel Johnson Chow, Lars Rex Christensen, Anthony Chu, Terry Chu, Vicki Chu, Kiwi Chui, Ryan Chun, Chung Kam Wah, Ricky Chung, Sharon Chung, Sarah Crang, Fiona M Deeley, Benjamond Diu, Du Quan, Feng Qiao Yu, Feng Yuan Lin, Maggie Feng, Lorna Fu, Wynne Fung, Gao Ming Rui, Anja Globke, Rowena Gonzales, Mark Robert Gridley, Joker Han, Kitty Han, Karen Hao, Har Wai Kin, Andrew Henderson, Eddy Ho, Maggie Ho, Nikki Ho, Rain Ho, Simon Ho, Kit Hong, Hou Yu, Sara Hu, Susen Hu, Colin Jiang, Jason Jiang, Unicorn Jiang, Wallace Jor, Apple Kam, Avni Kam, Anna Kao, Jofi Ko, Fian Kwan, Joanne Kwan, May Kwan, KK Kwok, Nicole Kwok, Phoenix Kwok, Tony Kwok, Vincent Kwok, Antonia Kwong, Nelson Kwong, Shirley Kwong, Andy Lai, Sandy Lai, David Lam, Watson Lam, Wing Lam, Christine Lau, Eric Lau, Erika Lau, Ernest Lau, Katy Lau, Matthew Lau, Ruby Lau, Tatum Lau, James Law, Ursula Law, Timothy Layton, Aidan Lee, Alvin Lee, Christie Lee, Janice Lee, Weldon Lee, Wendy Lee, Alice Lei, Bry Leung, Mei Ling Leung, Mike Leung, Monee Leung, Samuel Leung, Thomas Leung, Apple Li, Charlotte Li, David Li, Fion Li, Gary Li, Li Guo Feng, Tiger Li, Vera Li, Candy Liang, Cindy Ling, William Lim, Bob Liu, Carman Liu, Henry Liu, Liu Hong Bin, Liu Yuan Bo, Loretta Liu, Natasha Liu, Ting Liu, Wesley Liu, Adam Lo, Cathy Lo, Ivan Lo, John Lo, Anna Lu, Samuel Lui, Anna Luk, Abby Lung, Cindy Luo, Angel Mag, Wilson Mak, Mandy Meng, Carine Miu, Jeff Mok, Pan Mok, Surbhika Moodra, Seiki Mori, Mu Hai Tao, Mandy Ng, Matthew Ng, Yan Ng, Alice Ngan, Ruth Oliver, Geoffrey Poon, Po Poon, Meredith Proctor, Ren Zhi Sheng, Rock Ren, Idris Clwyd Roberts, Kido Ryoyu Andreas, Sang, Elaine Sha, Ling Shan, Sheng Hong Gang, Aska Shu, Julian Sit, Mavis Siu, Lynne Mary Smith, Edmond So, Sebastian Steinbock, Karen Elizabeth Stevens, Ada Sun, Jessis Sun, Nancy Sun, Sun Xin, Kinson Tam, Toki Tam, Carmen Tang, Joe Tang, Tang Chi Yuen, Teresa Tang, Tim Tang, Josephine Tao, Elaine Thomas, Teresa Tian, Tian Xiao Xu, Noel To, Eddie Tsang, Roy Tsang, Tony Tsang, Nick Tse, Tse Teak Seong, Clinton Tsoi, Raven Tsoi, Chris Tsui, Arthur Wai, Joey Wan, Terry Wan, Angela Wang, Enda Wang, Golden Wang, Lily Wang, Tiger Wang, Wang Lei, Wilson Wang, Aaron Wong, Benjamin Wong, Cecilia Wong, Edmond Wong, Frankie Wong, Heather Wong, Jodie Wong, Karie Wong, Lily Wong, Pak Wong, Wong Wing Hang, Yan Wong, Yokie Wong, Anya Wu, Lily Wu, Wu he ping, Bonny Xu, Crystal Xu, Xu Le, Xu Na, Xu Xiu Xiu, May Xue, Daniel Yang, Nicola Yau, Janice Yeung, Kelvin Yeung, Kenneth Yeung, Pat Man Yeung, Tommy Yeung, Rosita Yip, Yip Chak Lung, Yip Tak Leung, Karen Young, Mona Yu, Windy Yu, Cathy Yuan, Cindy Yuan, Francis Yuen, Jody Yuen, Jeff Yung, Yuri Vanini, Gordon Zeng, Zhai Ying Na, Coni Zhang, Leo Zhang, Zhang Chun Yan, Zhang Hai N, Anita Zhao, Meiko Zhao, Zhao Xin, Anne Zheng, Jason Zheng, Michael Zheng, Danae Zhou, Joe Zhou, Peter Zhou, George Zhu

图书在版编目（CIP）数据

旅：汉英对照 / 林伟而, 姜静, 南雪倩编著. --
北京：中国书籍出版社, 2012.5
 ISBN 978-7-5068-2794-2

Ⅰ. ①旅… Ⅱ. ①林… ②姜… ③南… Ⅲ. ①饭店—
建筑设计—作品集—中国—现代 Ⅳ. ①TU247.4
 中国版本图书馆CIP数据核字(2012)第078702号

《旅》

策　　划 ／	姜　静　南雪倩
特邀编辑 ／	刘　宏
责任编辑 ／	李国永　赵丽君
责任印制 ／	孙马飞　张智勇
封面设计 ／	思联建筑设计有限公司
出版发行 ／	中国书籍出版社
地　　址 ／	北京市丰台区三路居路97号(邮编：100073)
电　　话 ／	(010)52257142(总编室)　(010)52257153(发行部)
电子邮箱 ／	chinabp@vip.sina.com
经　　销 ／	全国新华书店
印　　刷 ／	北京瑞禾彩色印刷有限公司
开　　本 ／	260毫米×260毫米　1/12
印　　张 ／	32
字　　数 ／	110千字
版　　次 ／	2012年6月第1版　2012年6月第1次印刷
定　　价 ／	398元

版权所有　翻印必究

思联建筑设计有限公司

地　　址／香港湾仔港湾道2号香港艺术中心15楼

电　　话／(852) 2527-1931

传　　真／(852) 2529-8392

电　　邮／CL3@CL3.com

网　　址／www.CL3.com

INTERIOR DESIGN 出品